77

Fortschritte der Chemie organischer Naturstoffe

Progress in the Chemistry of Organic Natural Products

Founded by
L. Zechmeister

Edited by
W. Herz, H. Falk,
G. W. Kirby, R. E. Moore,
and Ch. Tamm

Authors:
W. A. Ayer, E. V. Brandt, J. Coetzee,
W. M. Daniewski, D. Ferreira, E. Malan,
L. S. Trifonov, and G. Vidari

Springer-Verlag Wien GmbH

Prof. W. Herz, Department of Chemistry,
The Florida State University, Tallahassee, Florida, U.S.A.

Prof. Dr. H. Falk, Institut für Chemie,
Johannes-Kepler-Universität, Linz, Austria

Prof. G. W. Kirby, Chemistry Department,
The University of Glasgow, Glasgow, Scotland

Prof. R. E. Moore, Department of Chemistry,
University of Hawaii at Manoa, Honolulu, Hawaii, U.S.A.

Prof. Dr. Ch. Tamm, Institut für Organische Chemie der Universität Basel,
Basel, Switzerland

Originally published by Springer-Verlag / Wien in 1999
Softcover reprint of the hardcover 1st edition 1999
Library of Congress Catalog Card Number AC 39-1015

Typesetting: Thomson Press (India) Ltd., New Delhi

Graphic design: Ecke Bonk
Printed on acid-free and chlorine-free bleached paper

SPIN: 10731467

With 3 partly coloured Figures

ISSN 0071-7886
ISBN 978-3-7091-7307-7 ISBN 978-3-7091-6366-5 (eBook)
DOI 10.1007/978-3-7091-6366-5

Contents

List of Contributors

AYER, Prof. W. A., Department of Chemistry, University of Alberta, Edmonton, Alberta, T6G 2G2, Canada

BRANDT, Prof. E. V., Department of Chemistry, University of the Orange Free State, P.O. Box 339, Bloemfontein 9300, South Africa

COETZEE, Dr. J., Department of Chemistry, University of the Orange Free State, P.O. Box 339, Bloemfontein 9300, South Africa

DANIEWSKI, Prof. W. M., Institute of Organic Chemistry, Polish Academy of Sciences, ul. Kasprzaka 44/52, 01-224 Warsaw, Poland

FERREIRA, Prof. D., National Center for the Development of Natural Products, Research Institute of Pharmaceutical Sciences, School of Pharmacy, The University of Mississippi, University, MS 38677, USA

MALAN, Prof. E., Department of Chemistry, University of the Orange Free State, P.O. Box 339, Bloemfontein 9300, South Africa

TRIFONOV, Dr. L. S., Department of Chemistry, University of Alberta, Edmonton, Alberta, T6G 2G2, Canada

VIDARI, Prof. G., Department of Chemistry, University of Pavia, Viale Taramelli 10, 27100 Pavia, Italy

Secondary Metabolites and the Control of Some Blue Stain and Decay Fungi

W. A. Ayer and L. S. Trifonov
Department of Chemistry, University of Alberta, Edmonton, Alberta, Canada

Contents

1. Introduction

Wood discoloration has been grouped in four categories based on the cause of the stain: enzymatic and chemical reactions, changes caused by contact with chemicals, chemical reactions accompanying the early stages of decay, and color changes associated with the growth of fungi (*1*). Only the latter two types of discoloration (biotic stains) will be discussed in this review. The discoloration by pigmented hyphae and spores on the wood surface is not included in this discussion.

Although blue stain fungi do not cause decay of wood, they lower the quality of wood and wood products (Fig. 1). In the pulping process, blue stain reduces pulp brightness, which leads to significant increases in the use of bleaching chemicals. It has been estimated that annual losses due to stain in the USA alone exceed $50 million (*1*).

The blue, black, brown or gray, or occasionally yellow, purple, pink or green discoloration in the sapwood is caused by the pigmented hyphae or by the release of pigments from the hyphae. Since the bluish discoloration is the most predominant, this stain is referred to as blue stain, sapstain, or mineral stain. It is believed that the discoloration is due to melanin-based pigments (*2*). In some cases the discoloration may be attributed to the formation of complexes of iron and other metals with phenolic compounds, especially with strongly chelating siderophores, which are produced by many fungi. It has been reported that the foliage of blue stain diseased lodgepole pine shows an increased level of iron with respect to foliage of healthy trees. This increase of iron uptake by plants caused by the accumulation of fungal iron chelators is considered as an iron stress response and may be linked to the characteristic discoloration of the sapwood of diseased trees (*3*).

In this review we will focus mainly on the chemical aspects of the blue stain and decay, especially on conifers and aspen wood.

Decay is the result of wood digestion (both lignin and cellulose) by fungi (Fig. 1). It is the major type of loss of wood which limits the use of

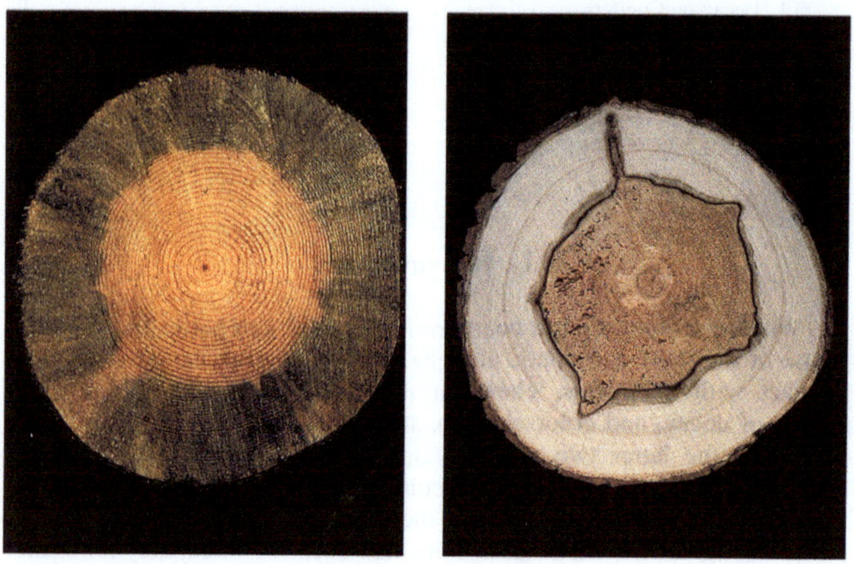

Fig. 1. Blue stained pine tree (left) and aspen decay (right)

conifer and broad leaved trees. There are two major types of decay, brown rots (mainly the carbohydrates of the cell wall are attacked by the fungi) and white rots (both carbohydrates and lignin are attacked by the fungi). The metabolites and control of both types of fungi will be discussed in this review.

Certain chemicals have been employed for preventing stain and decay. The application of solutions of sodium carbonate, borax, chlorinated phenols, organic mercury compounds, copper compounds, organic nitrogen containing heterocyclic or quaternary ammonium fungicides have been reported (*4*). The use of many of these preservatives, such as mercury compounds and pentachlorophenol, has been severely restricted because of environmental and worker's exposure concerns. Currently there is a growing trend to develop acceptable alternatives for chemical prevention of blue stain development. Progress has been made in biological control of stain fungi (*5*, *6*, *7*, *8*).

2. Blue Stain Fungi

2.1. Blue Stain on Conifers

The blue stain disease of conifers causes the death of more than 40 million trees a year in Western Canada where the mountain pine beetle is the vector of the disease. The microflora associated with the mountain pine beetle consists of several species of yeasts and mycelial fungi. Four species of the genus *Ceratocystis* have consistently been isolated from stained conifer wood: *C. clavigera*, *C. huntii*, *C. ips*, and *C. minor*. Fungi of this genus are also responsible for the oak wilt and Dutch elm diseases (*9*).

2,3-Dihydroxybenzoic acid (**1**), a well known iron-chelating agent (siderophore), has been identified amongst the secondary metabolites of some of these species (Chart 1). Compound **1** is involved in the iron transport systems of microbes via iron-chelating. It gives a bright blue complex with ferric ions (**2**) and greenish-brown complex with cobalt. In addition to a large number of ubiquitous metabolites, some simple phenolics, hydroxyisocoumarins (**3**), and hydroxydihydroisocoumarins (**4**) have been identified. Ceratenolone (**5**), isolated from *C. minor*, is one of the strongest fungal iron chelators (*10*). All these siderophores may be responsible for the staining effect, due to complexation with ferric and other ions (*3*, *9*, *10*).

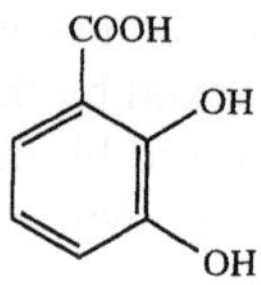

1

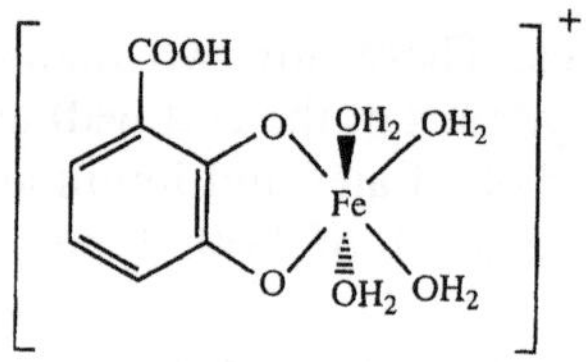

2

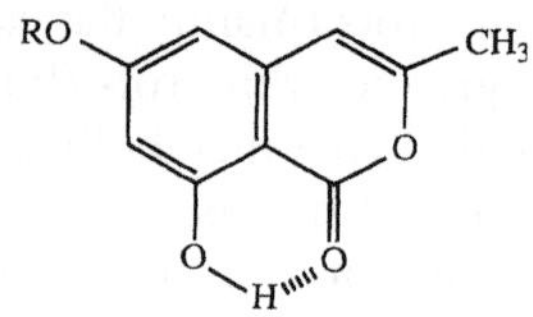

hydroxyisocoumarins (**3**)

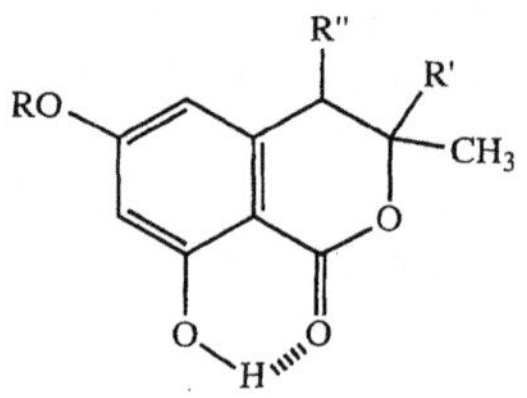

hydroxydihydroisocoumarins (**4**)

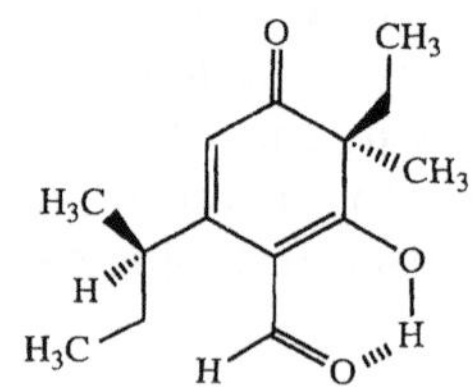

ceratenolone (**5**)

Chart 1. Siderophores isolated from *Ceratocystis* species

2.2. Blue Stain on Broadleaf Trees

Aspen (*Populus tremuloides*) is the most widely distributed tree species in North America. In Canada aspen represents more than 50% of the merchantable hardwood timber. Its previously neglected utilization has increased substantially in recent years. A serious limitation to its use in the pulping process, however, is the pronounced susceptibility to blue stain caused by fungi (*11*).

The two most important aspen blue stain fungi are *Ophiostoma crassivaginatum* and *O. piliferum*. The secondary metabolites of the blue stain fungus *O. crassivaginatum* were studied in an attempt to obtain insight into the nature of the staining material. Several simple phenolic compounds were identified which, at least in part, can be responsible for the discoloration of aspen wood infected with this fungus (*12*).

The aspen fungi *Peniophora polygonia* and *Sporormiella similis*, the former showing decay and the latter showing blue-stain antifungal activity, were studied for production of secondary metabolites. In both cases isobutyric and isovaleric acid were amongst the most active metabolites inhibiting the growth of decay and stain fungi (*5, 13*).

References, pp. 17–20

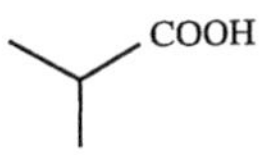

6

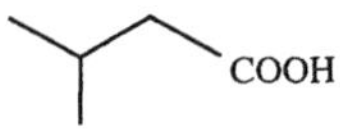

7

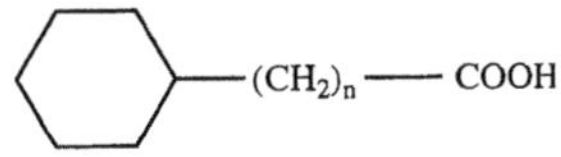

8 (n = 0-3)

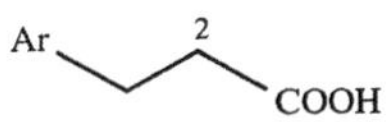

9
10 2(3)-en

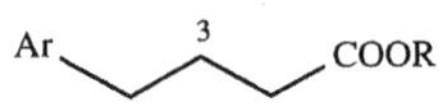

11
12 3(4)-en,.R = H

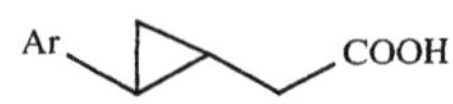

13

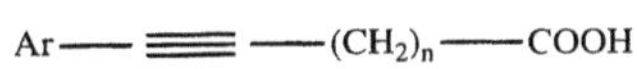

14 (n = 0, 1)

Chart 2. Some antifungal homologs and analogs of isobutyric acid

Following this natural lead, several homologs and analogs of isobutyric (**6**) and isovaleric acid (**7**), notably, the known enzyme inhibitor 4-phenyl-3-butenoic acid (**12**) were tested (Chart 2). Of 38 compounds tested, strong inhibitory activity, superior to that of some commercial fungicides, was observed for several cyclohexyl aliphatic acids (**8**) and for the enzyme inhibitor **12** and its triple bond analogs **14**, while the cyclopropane analogs **13**, the cinnamic acids **9** and **10**, and **11** proved less active (*14*). The acids **12** and **14** prevented colonization of aspen wood chips by the two stain fungi at concentrations as low as 1 μg/ml.

In a similar approach, senecioic acid (**15a**) and tiglic acid (**16**), minor, but very active anti-blue stain fungi compounds isolated from *S. similis*, were the natural leads (*5*) (Chart 3). The antifungal activities of 25 derivatives and analogs were studied. Senecioic acid, tiglic acid, 2,4-dihydroxyseneciophenone (**17a**) and the analogs cyclohexylidene acetic acid (**18**) and cyclohexene-1-carboxylic acid (**19**) completely prevented colonization of aspen wood chips at 10 μg/ml. Consequently, it was suggested that some of these active compounds are good candidates for chemical protection of aspen wood and wood products (*15*).

Stachybotris cylindrospora often has been isolated from clean xylem tissue of *P. tremuloides*. The metabolites produced by *S. cylindrospora* cultured in liquid medium were subjected to bioassay guided separation on XAD-16 non-ionic resin, followed by Sephadex LH-20 and silica gel chromatography to afford two known antifungal sesquiterpenes, trichodermin (**20**) and trichodermol (**21**) (Chart 4). Both compounds

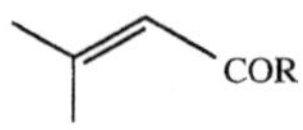

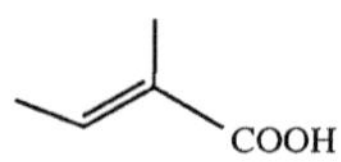

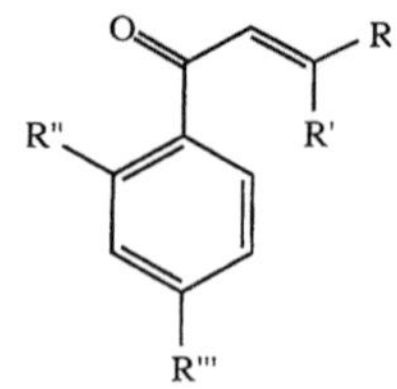

senecioic acid (**15a**: R = OH)

15b: R = OPh, NH_2, NHR', NHAr

tiglic acid (**16**)

seneciophenones (**17**)
(**17a:** R=R'=CH_3,
R''=R'''=OH)

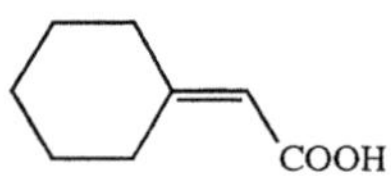

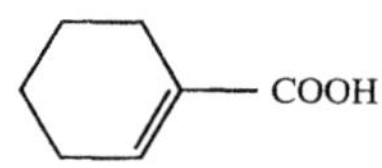

cyclohexylideneacetic acid (**18**)

cyclohexene-1-carboxylic acid (**19**)

Chart 3. Antifungal derivatives and analogs of senecioic and tiglic acid

significantly reduced the growth of the aspen blue stain fungus *O. crassivaginatum* at 10 μg/ml and completely inhibited the growth at 100 μg/ml. The novel sesquiterpene metabolites stachybotrydial (**22**) and stachybotramide (**23**), as well as the isochromane **24** showed no inhibitory activity. Based on the strong antagonism between *S. cylindrospora* and *O. crassivaginatum*, as well as on the innocuous nature of *S. cylindrospora*, it was suggested that this fungus could be used as a bioprotectant of *P. tremuloides* wood against staining (*6*).

An unidentified *Zythiostroma* species, isolated from aspen, was also found to be antagonistic to *O. crassivaginatum.* Zythiostromic acids A (**25**) and B (**26**), as well as zythiostromolide (**27**) isolated from the culture broth of this fungus, although new natural compounds, were inactive against the blue stain fungi (*16*) (Chart 5).

The fungus *Lecythophora hoffmannii* has been shown (*17*) to be strongly antagonistic to some blue stain fungi. Liquid cultures of the fungus show weak and variable activity against those blue stain fungi. The C-glucoside lecytoside (**28**) was isolated from the liquid culture but showed very weak activity *in vitro* against the target fungi (*18*) (Chart 6). However, when the fungus was grown on solid rice medium, a very active compound called lecythophorin was obtained (*19*). Extensive NMR studies, combined with some degradation studies, showed that lecythophorin possesses structure **29**, closely related to the known antibiotic chaetiacandin (**30**). In fact, beginning with the benzylic

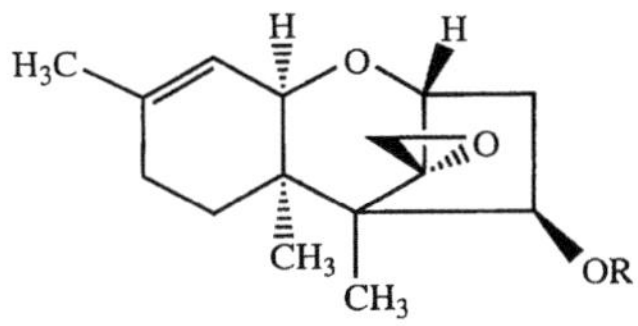

trichodermin (**20**), R = Ac
trichodermol (**21**), R = H

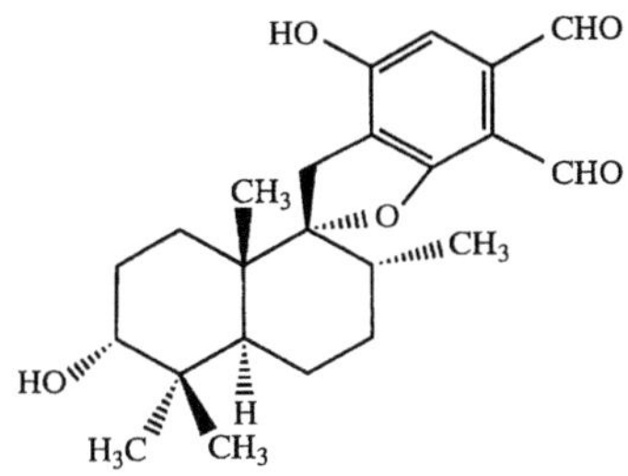

stachybotrydial (**22**)

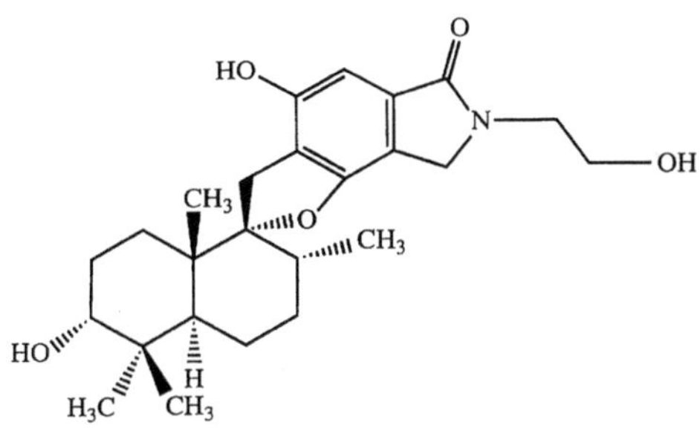

stachybotramide (**23**)

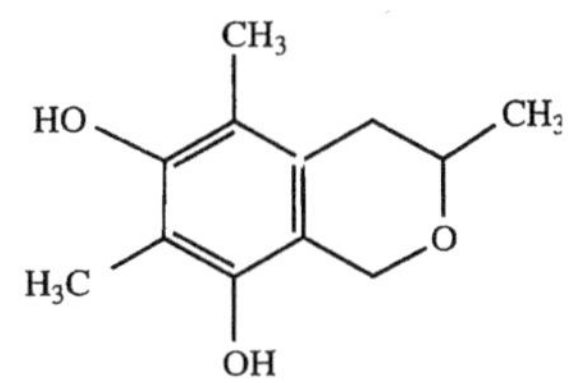

24

Chart 4. Secondary metabolites of *Stachybotrys cylindrospora*

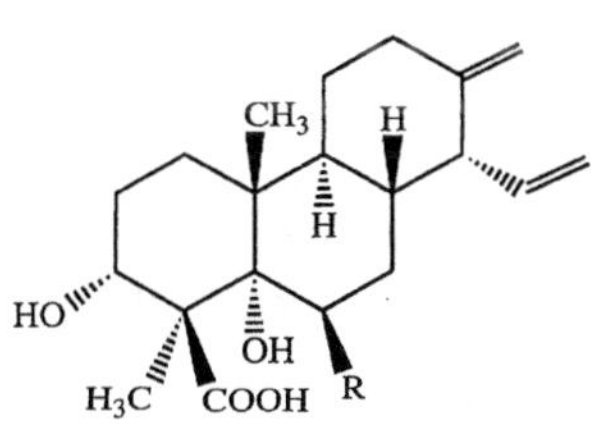

zythiostromic acid A (**25**), R = OH
zythiostromic acid B (**26**), R = H

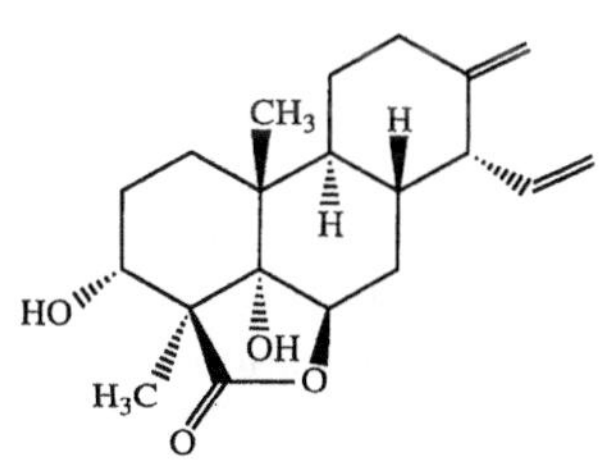

zythiostromolide (**27**)

Chart 5. Diterpenes from an unidentified *Zythiostroma* species

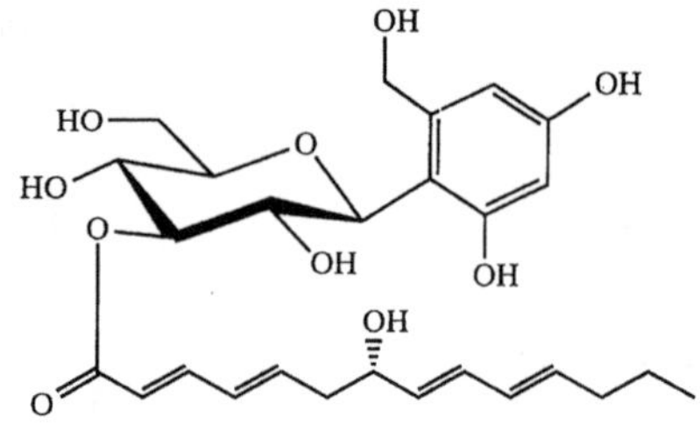

lecythoside (**28**)

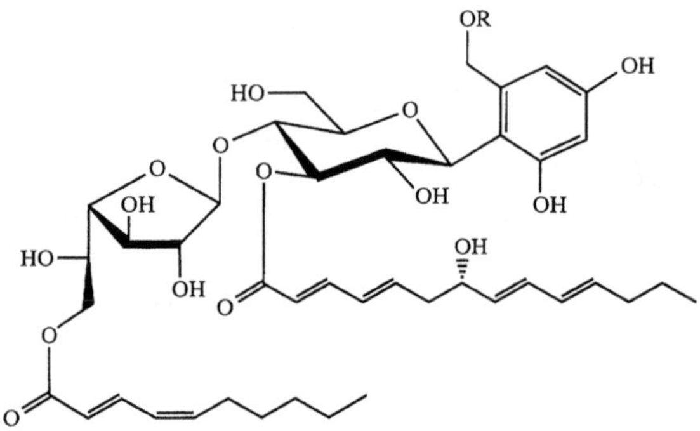

lecythophorin (**29**) R = SO_3Na

chaetiacandin (**30**) R = H

Chart 6. C-glucosides of *L. hoffmannii*

methylene group, it was possible to correlate all the C's and H's in the aromatic ring and the glucose and galactose portions of the structure. The presence of the galactose ring was verified by alkaline hydrolysis followed by methanolysis with HCl to give O-methylgalactopyranose. The furanoside ring, not unexpectedly, was transformed to the pyranoside by this operation.

Lecythophorin (**29**) shows very good activity at 1 μg/ml against the blue stain fungi (*20*), but the fact that the fermentation yield is very low and synthetic preparations of this compound will, presumably, not be possible on a commercial scale, precludes the use of this compound in the field.

3. Decay Causing Fungi

3.1. Decay on Conifers

A list of common decay fungi on coniferous trees is presented in Table 1 (*21*).

Anisomyces odoratus has been reported to produce odoriferous simple aromatic compounds, terpenes and lanostadienoic acids (*22*). The unusual amino acid 2-amino-4-N-ureidopropionic acid and its oxalyl derivative were recently isolated from *Coniophora puteana* (*23*).

Only steroids have been isolated from *Echinodontium tinctorium* and *Fomitopsis pini* (*22*). The long chain 2-hydroxy- and 2,3-dihydroxy fatty acids of ceramide, the n-phytosphingosines and the antitumor polysaccharides of *Fomitopsis pinicola* have been studied (*24, 25*). *Fomitopsis officinalis* yielded ergosterol-related steroids, officinalic acid, polyacetylenes and agaricic acid (*22*), as well as so-called agaric acid. The latter has been separated by TLC into three steroidal components (*26*).

A variety of benzoquinones, phenols and steroids have been isolated from the conifer decay fungus *Gloeophyllum saepiarium* (*22*).

Inonotus hispidus is a decay fungus rarely occurring on conifers but predominantly affecting deciduous trees. Recently it was reported that the ethanolic extract of fruiting bodies of *I. hispidus* contains a new pigment, hispolon (**31**) (Chart 7) which is accompanied by the known

Table 1. *Some Common Decay Fungi of Coniferous Trees*

Anisomyces odoratus (Wulf.:Fr.) Pat. = *Trametes odoratus* (Wulf.:Fr.) Fr.
Armillaria ostoyae (Romag.) Herink = *A. obscura* (Pers.) Herink
Coniophora puteana (Schum.:Fr.) Karst
Echinodontium tinctorium (Ell. & Ev.) Ell. & Ev. = *Fomes tinctorium* Ell. & Ev.
Fomitopsis officinalis (Vill.:Fr.) Bond. & Sing. = *Fomes officinalis* (Vill.:Fr.) Newman
Fomitopsis pinicola (Sw.:Fr.) Karst. = *Fomes pinicola* Sw.:Fr.
Gloeophyllum saepiarium (Wulf.:Fr.) Karst. = *Lenzites saepiarium* (Wulf.:Fr.) Fr.
Haematostereum sanguinolentum (Alb. & Schw.:Fr.) Pouzar = *Stereum sanguinolentum* Alb. & Schw.:Fr.
Hirschioporus abietinus (Dicks.:Fr.) Donk = *Polyporus abietinus* Dicks.:Fr.
Inonotus hispidus (Bull. ex Fr.) Karst.
Inonotus tomentosus (Fr.) Gilbertson = *Polyporus tomentosus* Fr.
Peniophora pseudo-pini Weres & S. Gibson
Phaeolus schweinitzii (Fr.) Pat. = *Polyporus schweinitzii* Fr.
Phellinus pini (Thore.:Fr.) Pil. = *Fomes pini* (Thore.:Fr.) Lloyd.
Pholiota alnicola (Fr.) Sing. = *Flammula alnicola* (Fr.) Quél.
Serpula himantioides (Fr.) Bond. = *Merulius himantioides* Fr.

styryl pyrone, hispidin. Both compounds have immunomodulatory and antiviral activity (*27*).

Phellinus pini is the most economically important conifer decay fungus which causes white pocket rot in many species of pine. Antagonism between *P. pini* and several other wood pathogens have been reported. When the fungus was grown in liquid medium several aromatic and benzoquinone metabolites were isolated of which 4-hydroxystyrene (**33**) and the pentaenone **35** showed activity against

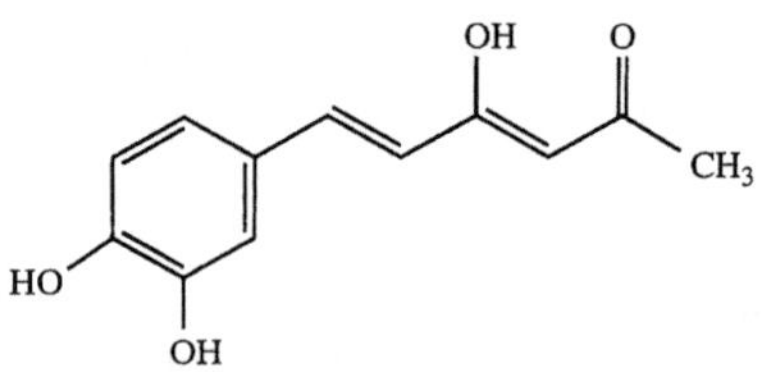

31

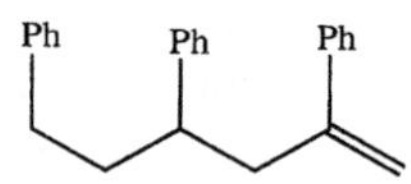

32

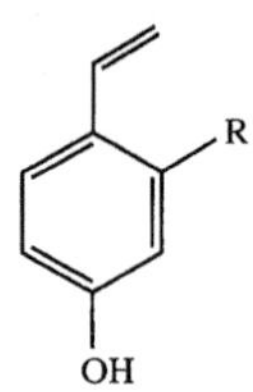

33, R = H
34, R = OH

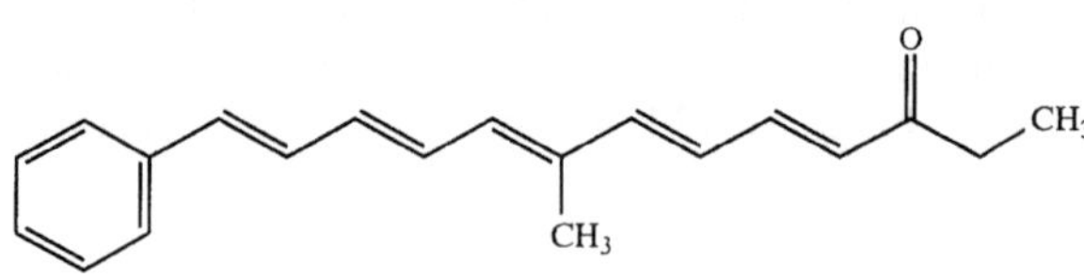

35

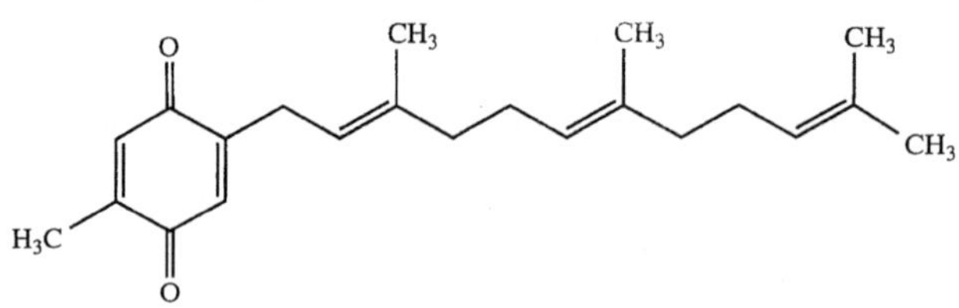

36

Chart 7. Secondary metabolites produced by *Inonotus hispidus* and *Phellinus pini*

References, pp. 17–20

several wood stain and decay fungi while compounds **32**, **34**, and **36** were inactive (*28*) (Chart 7).

The secondary metabolites of the remaining decay fungi listed in Table 1 have not been studied so far.

3.2. Decay on Broadleaf Trees

A list of common decay fungi on broadleaf trees is presented in Table 2 (*21*).

The production of aryl sesquiterpene metabolites (**37–39**) (Chart 8) by *Armillaria ostoyae*, a highly forest-pathogenic fungus, is enhanced up to 400-fold when the fungus is grown in the presence of an antagonist. Under these conditions several new metabolites are also produced. A detailed structure-antibiotic activity study of these metabolites has been conducted (*29, 30*). The odorous substancies of *Bjerkandera adusta* have been studied (*31*). The white rot fungus *Coriolus versicolor* (*32*) produces highly oxygenated cytotoxic steroids (**40**, **41**).

Fomes fomentarius is a parasitic fungus of beech trees from which several benzotropolone pigments (**42–45**) and an unusual dicarboxylic acid (**46**) have been reported (*33-36*) (Chart 9).

A series of highly oxygenated bitter triterpenoids, ganoderenic acids **47–51** and furanoganoderic acid (**52**) (Chart 10) have been found in

Table 2. *Some Common Decay Fungi of Broadleaf Trees*

Armillaria ostoyae (Romag.) Herink = *A. obscura*(Pers.) Herink
Bjerkandera adusta (Willd.:Fr.) Karst = *Polyporus adusta* Willd.:Fr.
Cerrena unicolor (Bull.:Fr.) Murr. = *Daedalea unicolor* Bull.:Fr.
Coriolus hirsutus (Wulf.:Fr.) Quél. = *Polyporus hirsutus* Wulf.:Fr.
Coriolus versicolor (L.:Fr.) Quél. = *Polyporus versicolor* L.:Fr.
Flammulina velutipes (Curt.:Fr.) Sing = *Collybia velutipes* Curt.:Fr.
Fomes fomentarius (L.:Fr.) Kickx = *Polyporus fomentarius* L.:Fr.
Ganoderma applanatum (Pers. ex Wallr.) Pat. = *Elfvingia applanatum* (Pers. ex Wallr.) Karst.
Gymnopilus spectabilis (Fr.) Sing. = *Pholiota spectabilis* (Fr.) Gill
Inonotus hispidus (Bull. ex Fr.) Karst.
Hirschioporus paragamenus (Fr.) Bond & Sing. = *Polyporus paragamenus* Fr.
Lyophyllum ulmarium (Bull.:Fr.) Kuehn. = *Pleurotus ulmarium* (Bull.:Fr.) Kumm.
Peniophora polygonia (Pers.Fr.) Bound = *Corticium polygonia* Pers.:Fr.
Phellinus tremulae (Bond.) Bond & Boriss. = *Fomes igniarius* (L.:Fr.) Kickx.
Pholiota destruens (Brond.) Quél.
Pholiota squarrosa (Pers.:Fr.) Kumm.
Piptoporus betulinus (Bull.:Fr.) Karst. = *Polyporus betulinus* Bull.:Fr.
Radulodon americanus Ryv. = *Radulum casearium* (Morg.) Lloyd

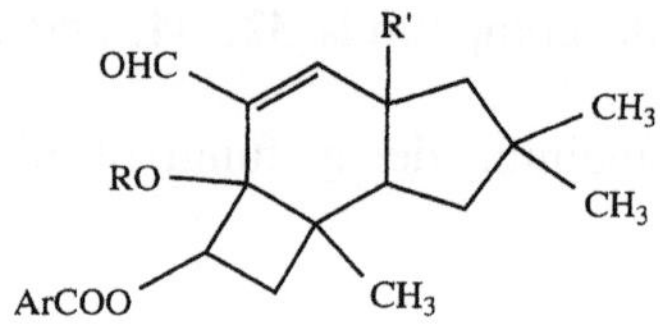

37, R = H or CH_3, R' = H or OH

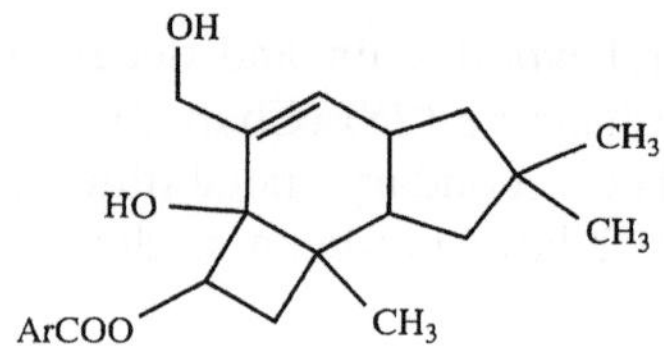

38

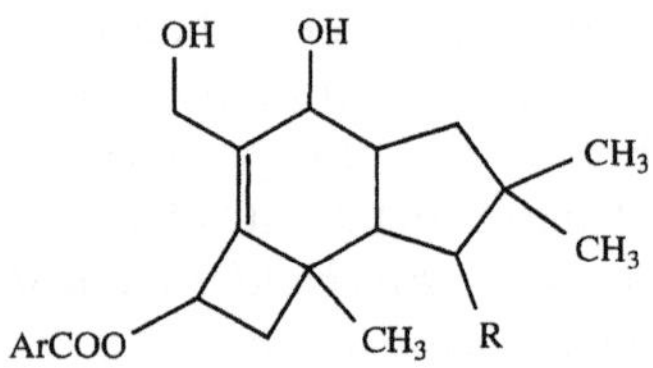

39, R = H or OH

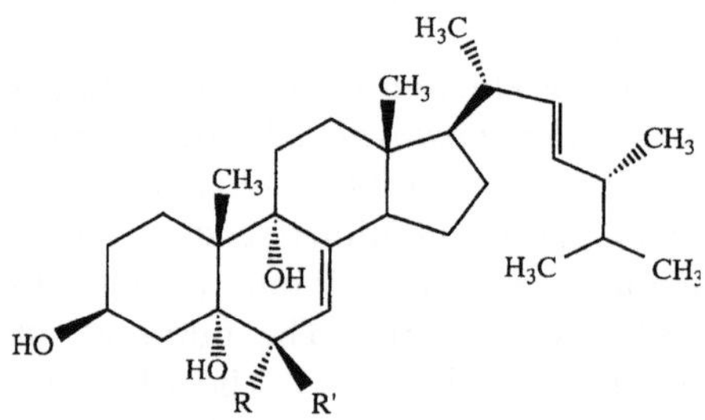

40, R + R' = O
41, R = H, R' = OCH_3

Chart 8. Sesquiterpene arylesters from *Armillaria ostoyae* (**37–39**) and steroids from *Coriolus versicolor* (**40–41**)

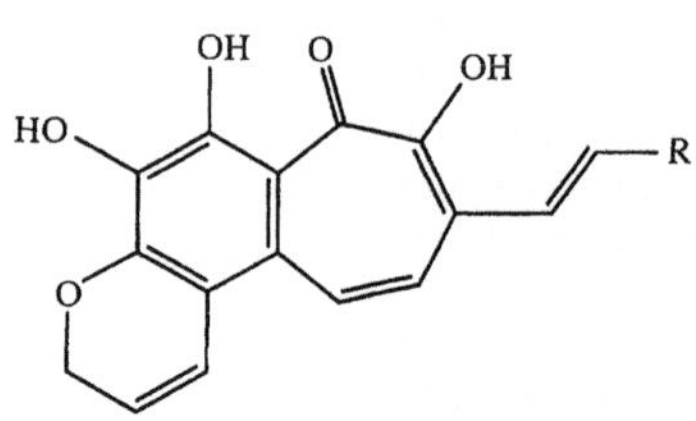

anhydrofomentariol (**42**), R = CH_2OH

anhydrodehydrofomentariol (**43**), R = CHO

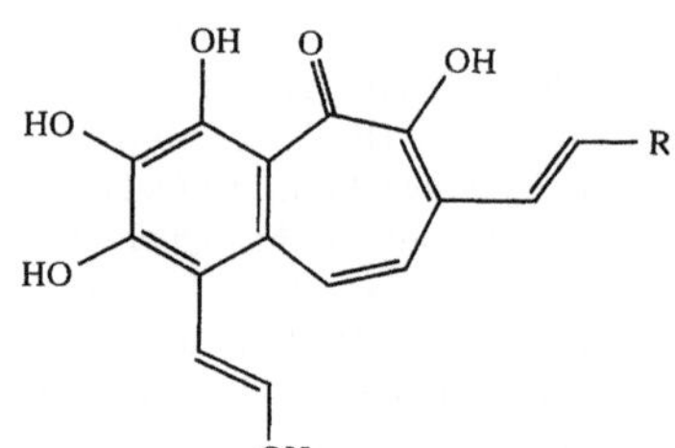

fomentariol (**44**), R = CH_2OH

dehydrofomentariol (**45**), R = CHO

$(C_{18}H_{37})_2C(COOH)CH(CH_3)COOH$

46

Chart 9. Benzotropolone pigments and a dicarboxylic acid from *Fomes fomentarius*

References, pp. 17–20

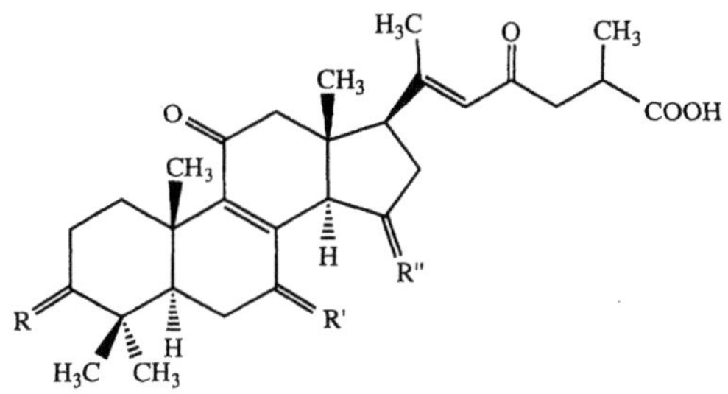

ganoderenic acid A (**47**): R = O, R' = β-OH, R'' = α-OH
ganoderenic acid F (**48**): R = R' = R'' = O
ganoderenic acid G (**49**): R = R' = O, R'' = α-OH
ganoderenic acid H (**50**): R = β-OH, R' = R'' = O
ganoderenic acid I (**51**): R = β-OH, R' = O, R'' = α-OH

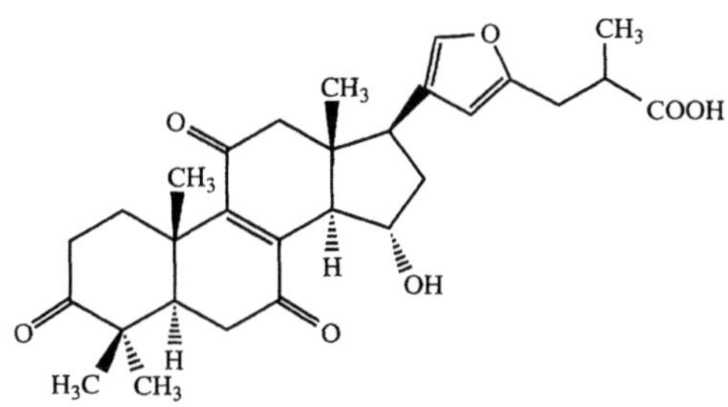

furanoganoderic acid (**52**)

Chart 10. Bitter triterpenes from *Ganoderma applanatum*

Ganoderma applanatum in addition to some other ergosterol-related steroids (*22, 37*).

Several polyacetylenes and phenolic compounds have been reported from *Gymnopilus spectabilis* (*22*) while the white rot fungus *Lyophillum ulmarium* produces unusual amino acids (**53**, **54**) and polyacetylenic compounds (**55–58**) (*38*, *39*) (Chart 11).

The most common decay fungi on aspen are *Phellinus tremulae* and *Peniophora polygonia*. The antagonism between *Phellinus tremulae* and *Peniophora polygonia* prompted a study of the secondary metabolism in search of the active fungal metabolites. In liquid culture *P. polygonia* produced a number of new drimane sesquiterpenes related to peniopholide (**59**), to cinnamolide (**63, 64**) or to confertifolin (**65**). In addition to the sesquiterpenes several new aromatic aldehydes (**66–69**)

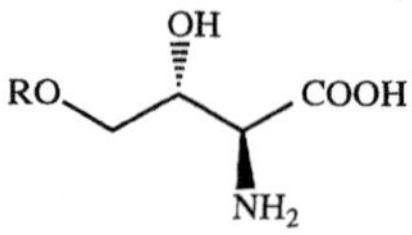

53, R = H
54, R = C_2H_5

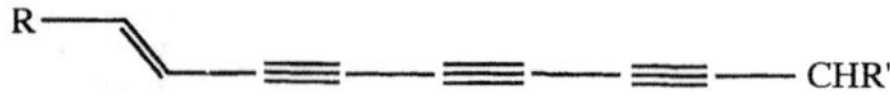

55, R = COOH, R' = OH
56, R = COOH, R' = H
57, R = CH_2OH, R' = H
58, R = CHO, R' = H

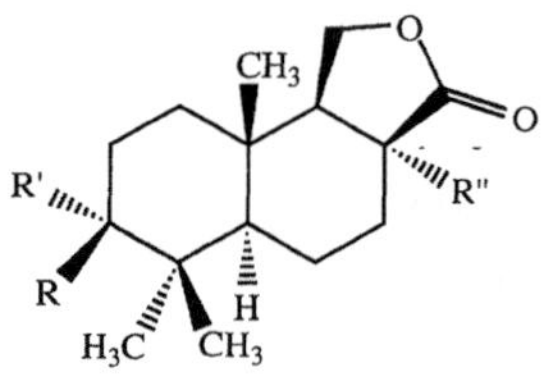

peniopholide (**59**), R = R' = H, R'' = OH
3α-hydroxypeniopholide (**60**), R = H, R' = OH, R'' = OH
3β-hydroxypeniopholide (**61**), R = OH, R' = H, R'' = OH
3β-hydroxydihydroconfertifolin (**62**), R = OH, R' = R'' = H

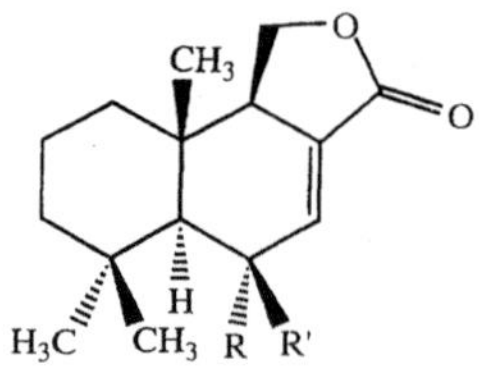

6α-hydroxycinnamolide (**63**), R = OH, R' =H
6β-hydroxycinnamolide (**64**), R = H, R' = OH

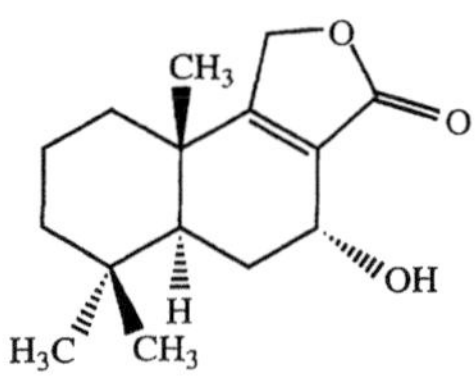

7α-hydroxyconfertifolin (**65**)

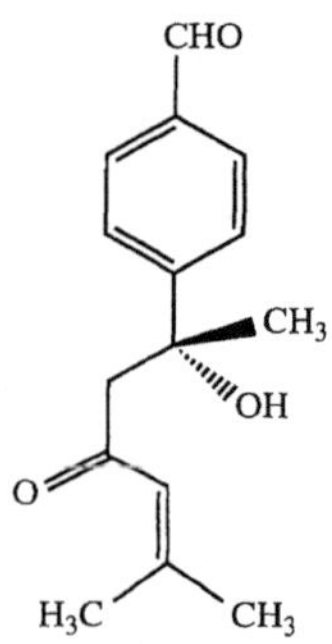

peniophoral (**66**)

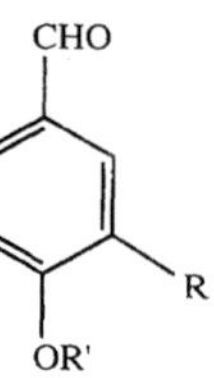

67, R=H, R'=$CH_2CH{=}C(CH_3)_2$
68, R=OH, R'=$CH_2CH{=}C(CH_3)_2$
69, R=H, R'=$CH_2CH{=}C(CH_3)CH_2OH$

Chart 11. Secondary metabolites from *Lyophillum ulmarium* (**53–58**) and *Peniophora polygonia* (**59–69**)

were identified (*40*, *41*) (Chart 11). Since none of these metabolites could account for the biological activity of *P. polygonia*, the volatile metabolites of this fungus were also examined. It was demonstrated that although they are minor volatile metabolites, senecioic acid and tiglic acid are very potent inhibitors of the growth of *Ph. tremulae* (*13*).

When grown in liquid medium *Phellinus tremulae* produced 2-carbomethoxyoxepin (**70**), the precursor of methyl salicylate, and a

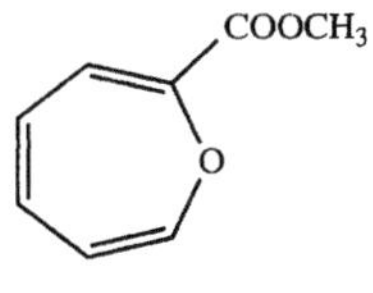

70

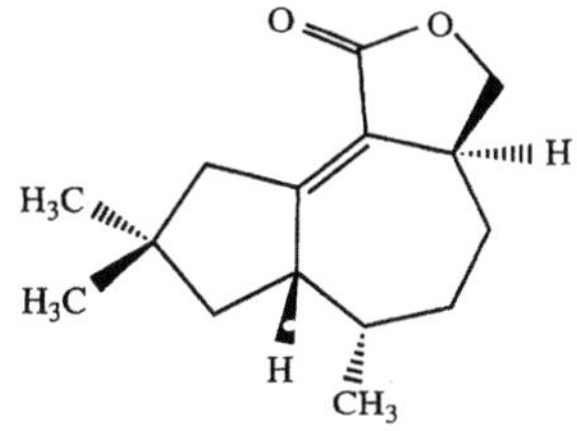

71

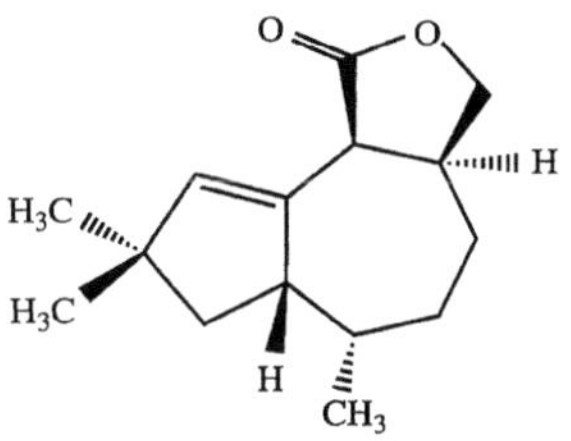

72

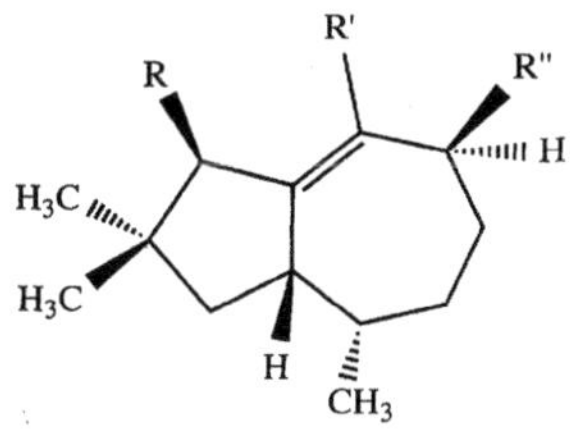

73, R = H, R' = R'' = CH_2OH
74, R = OH, R' = CH_3, R'' = CH_2OH

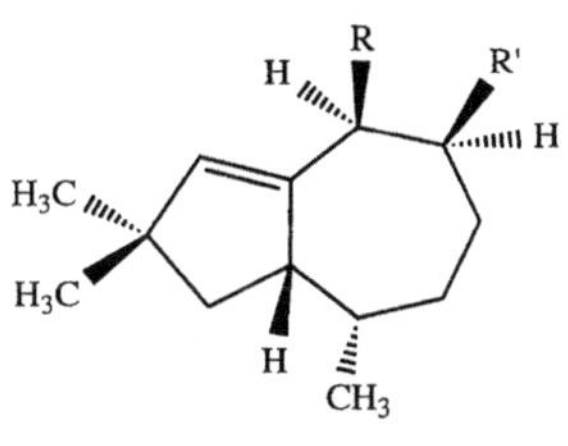

75, R = R' = CHO
76, R = R' = CH_2OH

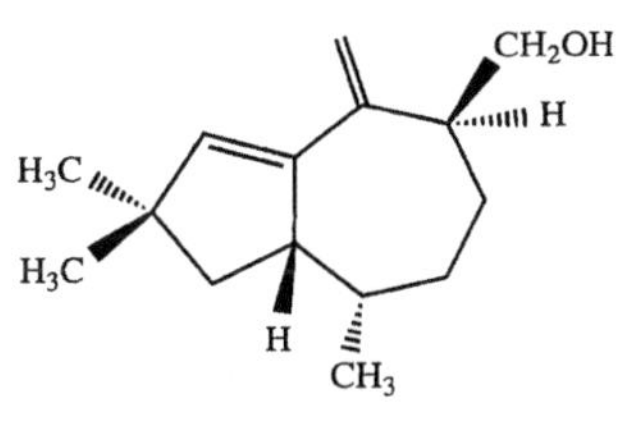

77

Chart 12. The metabolites of the aspen rotting fungus *Phellinus tremulae*

series of sesquiterpenes with a new perhydroazulene skeleton, named tremulanes (**71**–**77**) (*42*) (Chart 12). The multistep biosynthesis of the tremulanes via humulene involving two carbon migrations have been studied by ^{13}C labelling experiments (*43*).

The new fungus *Phoma etheridgei* isolated from black galls of aspen was known to be strongly antagonistic to the aspen decay fungus *Phellinus tremulae*. The bioassay guided isolation of metabolites from

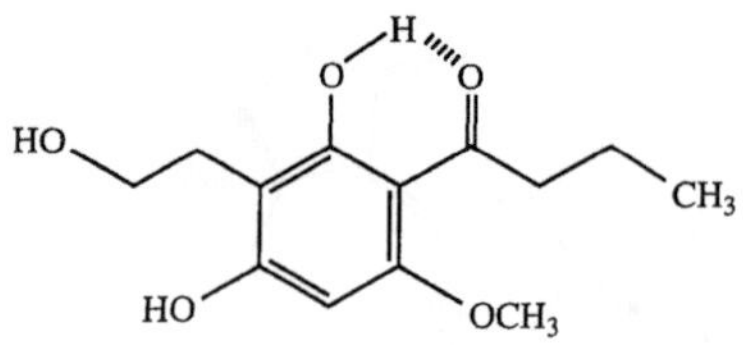

phomalone (**78**)

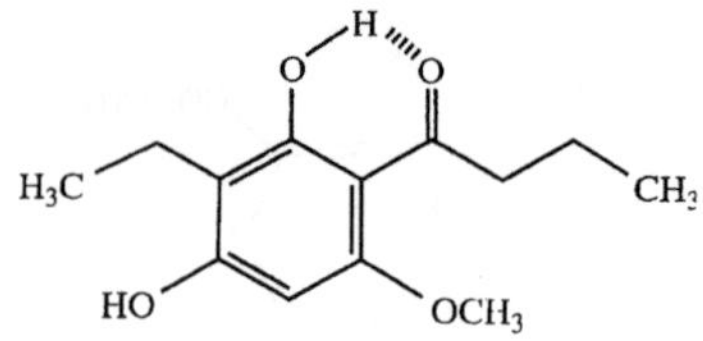

79

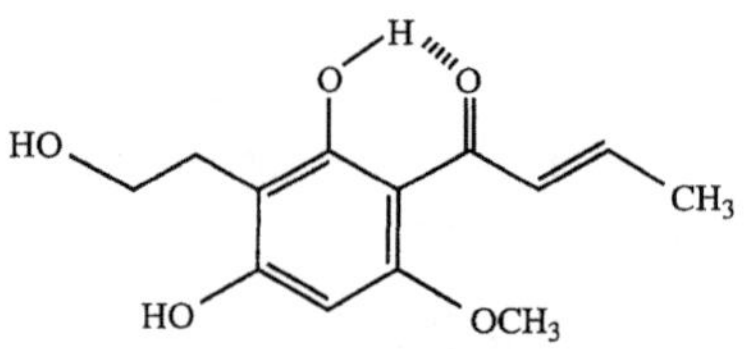

80

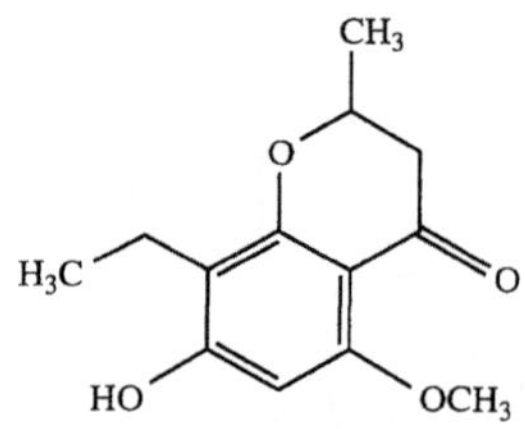

81

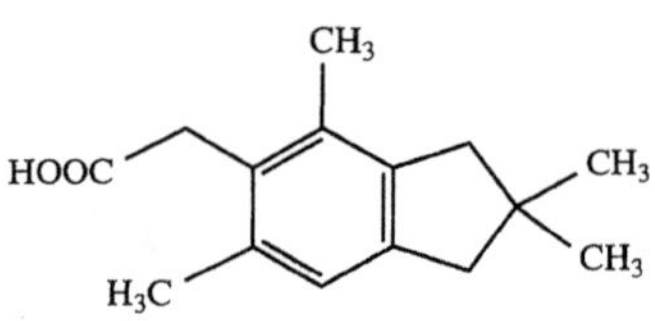

pholiotic acid (**82**)

Chart 13. Phenolic metabolites from *Phoma etheridgei* (**78–81**) and *Pholiota* destruens (**82**)

the liquid medium of this fungus led to the isolation of the new phenolic metabolites **78–81**, of which phomalone (**78**) was the most active *against P. tremulae* (Chart 13). ^{13}C-labelling experiments showed that the label of [2-^{13}C]-acetate is incorporated in both carbons of the ethyl side chain which is consistent with a biosynthetic way via a symmetrical cyclopropane intermediate (*44*).

Pholiota destruens, a white rot fungus, produces an antifungal and cytostatic illudalane sesquiterpene, pholiotic acid (**82**) (*45*), while *Piptoporus betulinus* was found to produce triterpenoid acids (*22*).

The secondary metabolites of the remaining fungi listed in Table 2 have not been studied so far.

4. The Black Gall Effect

It has been reported by HIRATSUKA that aspen trees with certain types of black galls occurring as a stem deformity have a lower incidence of decay caused by *Phellinus tremulae* than do nearby non-gall trees (*46*). A quantitative analysis of free benzoic and salicylic acid content of wood revealed that black gall tissue contained 1700 ppm of benzoic acid, the wood from the gall-bearing tree contained 90 ppm, and the wood from the non-gall tree contained only 14 ppm (*47*). The distribution of salicylic acid was similar. The high concentration of free acids was shown to be paralleled by the amounts of glycoside-bound benzoic and salicylic acids (*48*). Thus the gall tissue contained 12150 ppm of glycoside-bound salicylic acid and 17000 ppm of benzoic acid while the wood from a non-gall tree contained 3900 ppm of salicylic acid and 4000 ppm of benzoic acid. It was not clear, however, whether the black gall effect is due to direct fungicidal activity of these aromatic acids or to the so called systemic acquired resistance (SAR) induced by salicylic acid (*11*). Our bioassays on agar plates indicate that both acids have higher antifungal activity against the conifer and aspen stain fungi than they do against the aspen decay fungi (*49*).

Acknowledgements

We wish to thank Dr. P. Chakravarty for helpful discussions. The financial assistance of the Natural Sciences and Engineering Research Council of Canada in the form of a Strategic Grant is gratefully acknowledged.

References

1. ZABEL, R.A., and J.J. MORREL: Wood stains and discoloration. In: Wood Microbiology. Decay and its prevention. Acad. Press, San Diego, CA, 1992.

2. ZINK, P. and D. FENGEL: Studies on the colouring matter of blue-stain fungi. Part 3. Spectroscopic studies on fungal and synthetic melanins. Holzforschung **44**, 163 (1990).

3. AYER, W.A., B. KRATOCHVIL, E. ALLEN, L.M. BROWNE, C. DUFRESNE, D. FIGUEROLA, and A. SZENTE: The chemistry of the blue stain fungi. Part 2. Some essential metal levels of diseased and healthy lodgepole pine. Canad. J. Chem. **64**, 910 (1986).

4. TESCHKE, K., C. HERTZMAN, R.A. FENSKE, A. JIN, A. OSTRY, C. VAN NETTEN, and W. LEISS: A history of process and chemical changes for fungicide application in the Western Canadian lumber industry: What can we learn. Appl. Occup. Environ. Hyg. **9**, 984 (1994).
5. CHAKRAVARTY, P., L.S. TRIFONOV, L.J. HUTCHISON, Y. HIRATSUKA, and W.A. AYER: Role of *Sporormiella similis* as a potential bioprotectant of *Populus tremuloides* wood against the blue stain fungus *Ophiostoma piliferum*. Canad. J. For. Res. **24**, 2235 (1994).
6. HIRATSUKA, Y., P. CHAKRAVARTY, S. MIAO, and W.A. AYER: Potential for biological protection against blue stain in *Populus tremuloides* with a hyphomycetous fungus, *Stachybotrys cylindrospora*. Canad. J. For. Res. **24**, 174 (1994).
7. CROAN, S.C., and T.L. HIGHLY: Biological control of the blue stain fungus *Ceratocystis coerulescens* with fungal antagonists. Mat. and Org. **25**, 255 (1990).
8. SEIFERT, K.A., C. BREUIL, L. ROSSIGNOL, M. BEST, and J.N. SADDLER: Screening for microorganisms with the potential for biological control of sapstain on unseasoned lumber. Mat. and Org. **23**, 81 (1988).
9. AYER, W.A., L.M. BROWNE, M.C. FENG, H. ORSZANSKA, and H. SAEEDI-GHOMI: The chemistry of the blue stain fungi. Part. 1. Some metabolites of *Ceratocystis* species associated with mountain pine beetle infected lodgepole pine. Canad. J. Chem. **64**, 904 (1986).
10. AYER, W.A., S.K. ATTAH-POKU, L.M. BROWNE, and H. ORSZANSKA: The chemistry of the blue stain fungi. Part. 3. Some metabolites of *Ceratocystis minor* (Hedgcock) Hunt. Canad. J. Chem. **65**, 765 (1987).
11. AYER, W.A.: 1994 E.W.R. Steacie Award Lecture. Application of natural products chemistry to a biological problem. Canad. J. Chem. **73**, 465 (1995).
12. AYER, W.A., and L.S. TRIFONOV: Phenolic and polyketide metabolites of the aspen blue-stain fungus *Ophiostoma crassivaginata*. Phytochem. **38**, 371 (1995).
13. TRIFONOV, L.S., P. CHAKRAVARTY, Y. HIRATSUKA, and W.A. AYER: Antifugal activity of metabolites of *Peniophora polygonia* against the aspen decay fungus *Phellinus tremulae*. Eur. J. For. Path. **22**, 441 (1992).
14. TRIFONOV, L.S., W.A. AYER, P. CHAKRAVARTY, and Y. HIRATSUKA: Structure-activity relationships of some congeners of 4-phenyl-3-butenoic acid against decay and stain fungi. Eur. J. For. Path. **26**, 273 (1996).
15. CHAKRAVARTY, P., Y. HIRATSUKA, L.S. TRIFONOV, and W.A. AYER: Antifungal activity of derivatives of senecioic and tiglic acids against three aspen pathogens. J. Plant Dis. Protection **104**, 254 (1997).
16. AYER, W.A., and A.Q. KHAN: Zythiostromic acids, diterpenoids from an antifungal Zythiostroma species associated with aspen. Phytochem. **42**, 1647 (1996).
17. CHAKRAVARTY, P., and Y. HIRATSUKA: Evaluation of *Lecythophora hoffmannii* as a potential biological control against a blue stain fungus on *Populus tremuloides*. Z. Pflanzenk. Pflanzens. **101**, 74 (1994).
18. AYER, W.A., E.R. CRUZ, N. KAWAHARA, D.J. MUIR, and K. PTASZINSKA: Chemistry associated with black galls on aspen. The metabolites of *Lecythophora hoffmannii*, a fungus strongly antagonistic to blue stain fungi. Revista Latinoamericana de Quimica **24**, 183 (1996).
19. AYER, W.A., and N. KAWAHARA: Lecythophorin, a potent inhibitor of blue-stain fungi, from the hyphomycetous fungus *Lecythophora hoffmannii*. Tetrahedron Letters **36**, 7953 (1996).

20. Chakravarty, P., and Y. Hiratsuka: Effect of lecythophorin on in vitro growth of three aspen pathogens and its role in controlling stain and decay of wood. Z. Pflanzenk. Pflanzens. **103**, 164 (1996).
21. Hiratsuka, Y.: Forest diseases of the prairie provinces. Information report NOR-X-286, Northern Forestry Centre, Canadian Forestry Service (1987).
22. Turner, W.B., and D.C. Aldridge: Fungal Metabolites, 2nd Ed. New York: Academic Press. 1983.
23. Evans, C.S., P.J. Burns, M. Dutton, S.Brown: 2-Amino-4-N-ureidopropionic acid (albizzine) and its oxalyl derivative in hyphae of *Coniophora puteana*. Phytochem. **29**, 2159 (1990).
24. Asawa, K., and T. Yoshimoto: Constituents of mushrooms. 6. The constituent fatty acids and long chain bases of ceramide from *Fomitopsis pinicola*. Mokuzai Gakkaishi **26**, 819 (1980); CA 94:117774n.
25. Mizuno, T.: Sarunokoshikake: Polyporaecea fungi — kofukisarunokoshikake, *Ganoderma applanatum* and tsugasarunokoshikake, *Fomitopsis pinicola*. Food Rev.Int. **11**, 129 (1995).
26. Schulte, K.E., G. Rücker, H. Fachmann: Zur Struktur der "Agaricolsäure". Tetrahedron Letters **48**, 4823 (1967).
27. Ali, N.A.A., R. Jansen, H. Pilgrim, K. Liberra, U. Lindequist: Hispolon, a yellow pigment from *Inonotus hispidus*. Phytochem. **41**, 927 (1996).
28. Ayer, W.A., D.J. Muir, and P. Chakravarty: Phenolic and other metabolites of *Phellinus pini*, a fungus pathogenic to pine. Phytochem. **42**, 1321 (1996).
29. Peipp, H., and J. Sonnenbichler: Occurrence of antibiotics in cultures of *Armillaria ostoyae* growing in the presence of an antagonistic fungus or host plant cells. Biol. Chem. Hoppe-Seyler, **373**, 675 (1992).
30. Sonnenbichler, J., J. Dietrich, and H. Peipp: Secondary fungal metabolites and their biological activities. V. Investigations concerning the induction of the biosynthesis of toxic secondary metabolites in Basidiomycetes. Biol. Chem. Hoppe-Seyler, **375**, 71 (1994).
31. Berger, R.G., K. Neuhauser, F. Drawert: Characterization of odour principles of some Basidiomycetes: *Bjerkandera adusta, Poria aurea, Tyromyces sambucens*. Flavour Fragrance J. **1**, 181 (1986).
32. Valisolalao, J., B. Luu, and G. Ourisson: Chemical and biochemical study of Chinese drugs. VIII. Cytotoxic steroids from *Polyporus versicolor*. Tetrahedron **39**, 2779 (1983).
33. Arpin, N., J. Favre-Bouvin, and W. Steglich: Chemotaxonomic research on fungi. XXVII. Fomentariol. New benzotropolone isolated from *Fomes fomentarius*. Phytochem. **13**, 1949 (1974).
34. Gripenberg, J., L. Hiltunen, and L. Niinisto: Dehydrofomentariol, $C_{17}H_{14}O_7$. Cryst. Struct. Commun. **5**, 571 (1976).
35. Favre-Bouvin, J., M. Kaouadji, and N. Arpin: Chemotaxonomic studies on the mushrooms. 35. Anhydrofomentariol, a new pigment from *Fomes fomentarius*. Phytochem. **16**, 495 (1977).
36. Favre-Bouvin, J., N. Arpin, and J. Gripenberg: Anhydrodehydrofomentariol, a new pigment from *Fomes fomentarius*. Phytochem. **16**, 1852 (1977).
37. Nishitoba, T., S. Goto, H. Sato, and S. Sakamura: Bitter triterpenoids from the fungus *Ganoderma applanatum*. Phytochem. **28**, 193 (1988).
38. Ogawa, T., Y. Oka, and K. Sasaoka: D_s-*erythro*-2-amino-4-ethoxy-3-hydroxybutanoic acid from the fruiting bodies of the edible mushroom *Lyophillim ulmarium*. Phytochem. **24**, 1837 (1985).

39. GARDNER, J.N., E.R.H. JONES, P.R. LEEMING, J.S. STEPHENSON: Chemistry of the higher fungi. X. Further polyacetylenic derivatives of decane from various basidiomycetes. J. Chem. Soc. 691 (1960).

40. AYER, W.A., and L.S. TRIFONOV: Drimane sesquiterpene lactones from *Peniophora polygonia*. J. Nat. Prod. **55**, 1454 (1992).

41. AYER, W.A., and L.S. TRIFONOV: Metabolites of *Peniophora polygonia*. Part 2. Some aromatic compounds. J. Nat. Prod. **56**, 85 (1993).

42. AYER, W.A., and E.R. CRUZ: The tremulanes, a new group of sesquiterpenes from the aspen rotting fungus *Phellinus tremulae*. J. Org. Chem. **58**, 7529 (1993).

43. CRUZ, E.R.: The biosynthesis of the new tremulene sesquiterpenes isolated from *Phellinus tremulae*. Can. J. Chem. **75**, 834 (1997).

44. AYER, W.A., and L.D. JIMENEZ: Phomalone, an antifungal metabolite of *Phoma etheridgei*. Canad. J. Chem. **72**, 2326 (1994).

45. BECKER, U., T. ANKE, and O. STERNER: A novel bioactive illudalane sesquiterpene from the fungus *Pholiota destruens*. Nat. Prod. Lett. **5**, 171 (1994).

46. CRANE, P.E., P.V. BLENIS, and Y. HIRATSUKA: Black stem galls on aspen and their relationship to decay by *Phellinus tremulae*. Canad. J. For. Res. **24**, 2240 (1994).

47. PAUSLER, M.G., W.A. AYER, and Y. HIRATSUKA: Benzoic acid, salicylic acid, and the role of black galls on aspen in protection against decay. Canad. J. For. Res. **25**, 1749 (1995).

48. AYER, W.A., and G. HERGER, unpublished results.

49. CHAKRAVARTY, P., and Y. HIRATSUKA, personal communication.

(Received October 19, 1998)

Condensed Tannins

D. Ferreira [1,*], E. V. Brandt [2], J. Coetzee [2], and E. Malan [2]

[1] National Center for the Development of Natural Products, Research Institute of Pharmaceutical Sciences, School of Pharmacy, The University of Mississippi, University, MS, USA

[2] Department of Chemistry, University of the Orange Free State, Bloemfontein, South Africa

Contents

1. Introduction

The condensed tannins (syn. polymeric proanthocyanidins) represent a major group of phenolic compounds in woody and some herbaceous plants (*1–3*). Their exceptional concentrations in the barks and heartwoods of a variety of tree species have resulted in their commercial extraction with the initial objective of applying the extracts in leather manufacture (*4*). Essentially all of their biological significance, *e.g.* the protection of plants from insects, diseases and herbivores, and most of the current, *e.g.* leather manufacture, and also most promising new uses, *e.g.* pharmaceuticals or wood preservatives, rest on their complexation with other biopolymers like proteins and carbohydrates, or metal ions (*5*, *6*). Increasing attention has thus been directed to understanding their conformation and conformational flexibility (*7–20*) in order to explain their biological activity and to provide a basis for further development of uses for these renewable phenolic compounds.

Recent developments have also been initiated by the growing realization that the condensed tannins may additionally be credited for the profound health-beneficial properties of tea, fruit juices and red wine. This is mainly due to their *in vitro* radical scavenging (*21*) or antioxidant (*22*) biological properties, while the polymeric proanthocyanidins in red wine have been implicated in protection against cardiovascular disorders (*23*), *e.g.* the "French paradox" (*24–26*).

Collectively these 'positive' characteristics of the polymeric proanthocyanidins have transformed "a relatively unattractive and therefore neglected area of study" (*27*) into, yet again, a fashionable research field. The past 20–25 years have thus witnessed remarkable growth in our understanding of the basic structures of these compounds (*1–3*, *28*, *29*). Results relevant to some of the recent developments in the chemistry of the condensed tannins constitute the subject of this review.

2. Formation of the Interflavanyl Bond in Oligomeric Proanthocyanidins

The aptitude to utilize this important class of biomolecules to the maximum benefit of mankind depends on the accurate knowledge of their structures. Owing to the adverse effects of conformational mobility

References, pp. 59–67

of both the interflavanyl bond(s) (*30–34*) and heterocyclic ring(s) (*35, 36*, see also refs. *7–20*) of the proanthocyanidin oligomers on ^{1}H and ^{13}C NMR spectra and due to the heterogeneity of absolute configurations of constituent flavanyl units, synthesis is often the only means to unequivocally establish structure. The limited number of options that are available for the synthesis of proanthocyanidin oligomers were comprehensively reviewed (*4, 27–29*) and do not deserve detailed discussion here.

The oligo- and polymeric proanthocyanidins usually originate *via* coupling at C(4) (C-ring) of an electrophilic flavanyl unit, generated from a flavan-4-ol or a flavan-3,4-diol (the chain extender units), to a nucleophilic flavanyl unit, most often a flavan-3-ol (*1, 3*) (the chain terminating unit). In principle, the chain extender flavanyl unit with a good leaving group (typically OH or SCH_2Ph) at C(4), *e.g.* (**1**) and (**2**), is converted into a strongly electrophilic centre, typically a C(4)-carbocation (**3**) or an A-ring *p*-quinone methide (**4**), which is then trapped by the chain terminating nucleophilic flavan-3-ol (**5**) to give the regiomeric (4→6)- and (4→8)-biflavanoids (**6**) and (**7**). These then serve as precursors to triflavanoids (**8**) *via* coupling with the electrophile at the A- or D-rings, and eventually to tetraflavanoids and higher oligomers (Scheme 1) (*28, 29*). In the profisetinidins, *e.g.* (**7**) (R^1=H), *i.e.* representative of the 5-deoxy (A-ring) series of proanthocyanidins, the biflavanoids display a preference for D-ring coupling in the transformation to the angular trimers, while the procyanidin-type biflavanoids, *e.g.* (**7**) (R^1=OH), *i.e.* representative of the 5-oxy (A-ring) series, exhibit a remarkable propensity for coupling at C(8) of the A-ring and hence for the formation of linear triflavanoids. For chain terminating moieties possessing B-rings with nucleophilicity comparable to that of the A-ring, the former rings often compete favourably as nucleophiles in the process of interflavanyl bond formation (*37, 38*).

The factors that control the feasibility and the stereochemical course of the coupling process, as well as the methods to establish the configuration at C(4) of the condensation products and the mode of interflavanyl linkage were sufficiently reviewed (*4, 27–29*). Acid-catalyzed reactions to produce flavan-4-carbocations or A-ring quinone methides that may react with the A-rings of flavan-3-ols to produce oligo- and polymeric proanthocyanidins have been so successfully employed that they were called "biomimetic syntheses" (*39, 40*). The most recent variations of this theme are now briefly discussed. The nomenclature delineated in ref. (*1*) will be consistently employed.

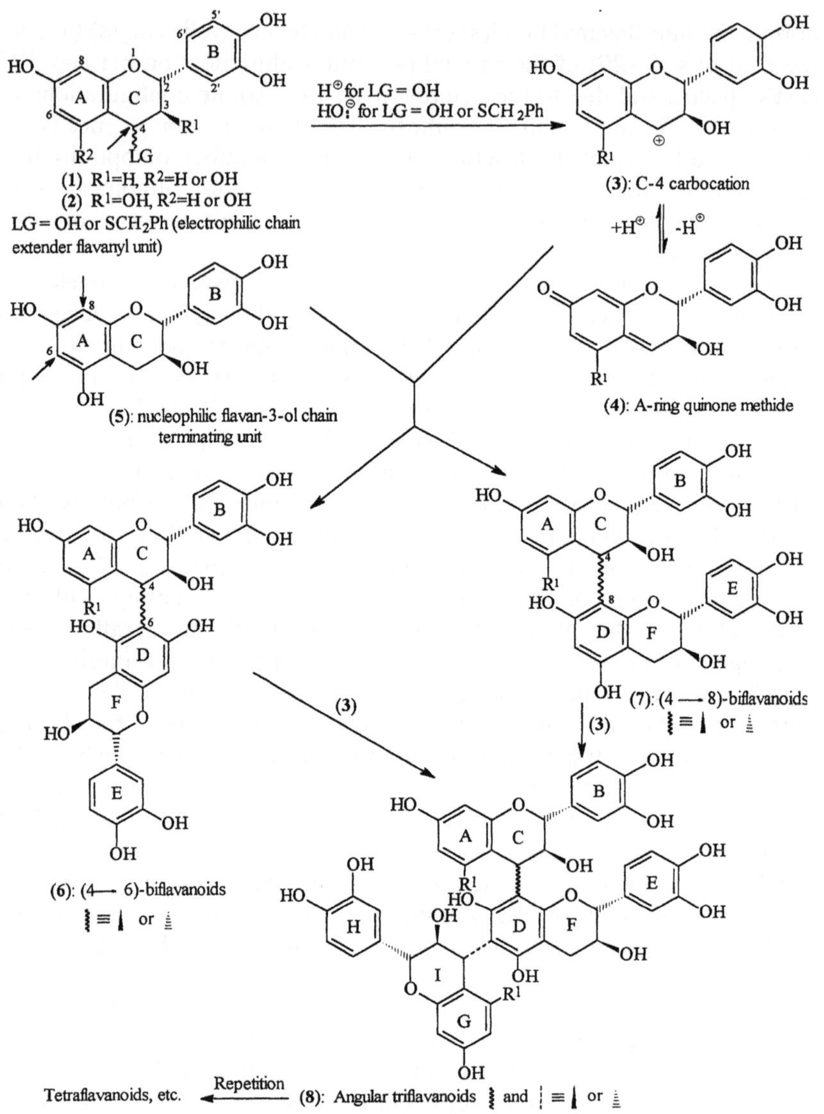

Scheme 1. Flavan-4-ols (**1**), flavan-3,4-diols (**2**) or 4-benzylsulfanyl analogues, and flavan-3-ols (**5**) as precursors to profisetinidin (R^1=H) or procyanidin (R^1=OH) oligomers with (2*R*,3*S*) constituent flavanyl units

2.1. Synthesis of the First Profisetinidins with Epifisetinidol Constituent Units

The profisetinidins, with their 7,3′,4′-trihydroxyflavan-3-ol chain extender units, are the most important polyflavanoids of commerce, forming the major constituents of wattle and quebracho tannins (see ref. (*41*)) for appropriate references). Naturally occurring oligomers exhibit predominantly 2,3-*trans* relative stereochemistry and possess either 2*R*,3*S* or 2*S*,3*R* absolute configurations (*4*, *27*). 5-Deoxy (A-ring) proanthocyanidins exhibiting 2,3-*cis* relative configuration of the chain extender moieties are extremely rare and are hitherto restricted to two tentative (4→6)-bis-fisetinidols (**9**) and (**10**) from *Colophospermum mopane* (*42*), a promelacacinidin (**11**) from *Acacia melanoxylon* (*43*) and two proteracacinidins (**12**) and (**13**) from *Acacia galpinii* (*44*).

The bark of *Pithecellobium dulce* (Roxb.) Benth (Guamúchil, Madras thorn) contains a rich array of mono-, di- and trimeric profisetinidins exhibiting both 2,3-*trans*- and 2,3-*cis* relative configuration of the constituent fisetinidol units (*41*), hence offering the first opportunity to rigorously corroborate the structures of the 2,3-*cis* analogues *via* synthesis. The synthetic approach was additionally motivated by the precariousness of unequivocally differentiating between 2,3-*cis*-3,4-*trans*- and 2,3-*cis*-3,4-*cis* configurations of chain extender units on the basis of ^{1}H NMR coupling constants (*45*, *46*). Furthermore, the powerful NOE method for differentiating between 2,4-*cis*- and 2,4-*trans*-substituents (*47*) is less useful at the di- and trimeric level due to the adverse effects of dynamic rotational isomerism about the interflavanyl bond(s) on ^{1}H NMR spectra at ambient temperatures (*30–33*).

Separate treatment of epifisetinidol-4β-ol (**14**) with catechin (**5**) and epicatechin (**15**) under mild acidic conditions afforded the epifisetinidol-(4β→6)- and (4β→8)-catechins (**16**) and (**18**), and epifisetinidol-(4β→6)- and (4β→8)-epicatechins (**17**) and (**19**), respectively (Scheme 2) (*41*), both couplings proceeding highly stereoselectively as anticipated (*40*). ^{1}H NMR data of the permethylaryl ether diacetates in conjunction with the chiroptical properties (*46*, *48*, *49*), based on the aromatic quadrant rule (*50*) permitted unambiguous structural confirmation of the natural products (**18**) and (**19**). Acid-catalyzed condensation of epifisetinidol-4β-ol (**14**) with either catechin (**5**) or epicatechin (**15**) in a 1:6 molar ratio stereoselectively afforded both the dimeric profisetinidins (**18**) and (**19**) and the angular trimeric profisetinidins (**20**) and (**21**). The appropriate derivatizations gave the permethylaryl ether triacetates with ^{1}H NMR and CD data identical to those of the corresponding derivatives of the natural products.

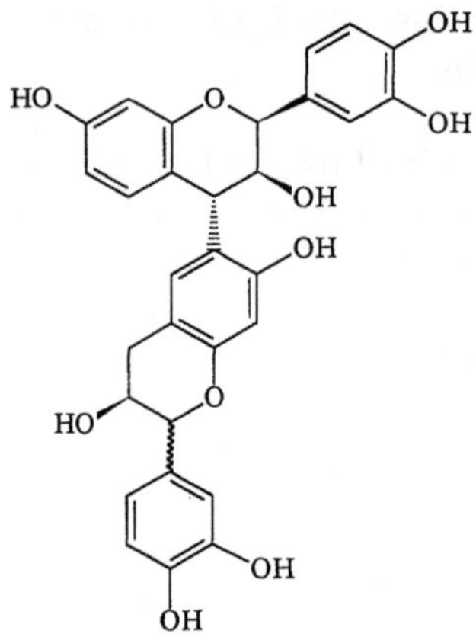

(9)
(10)

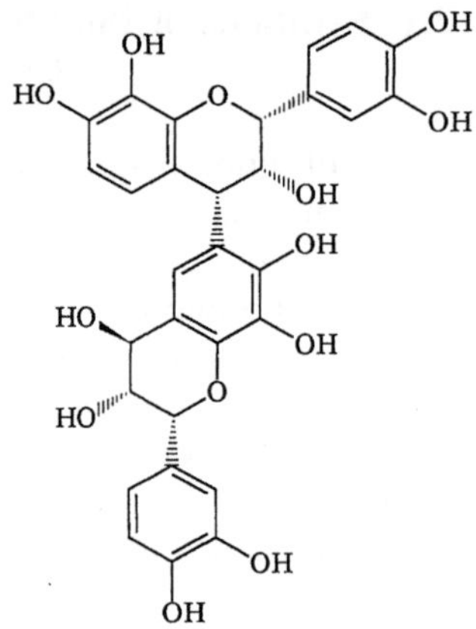

(11)

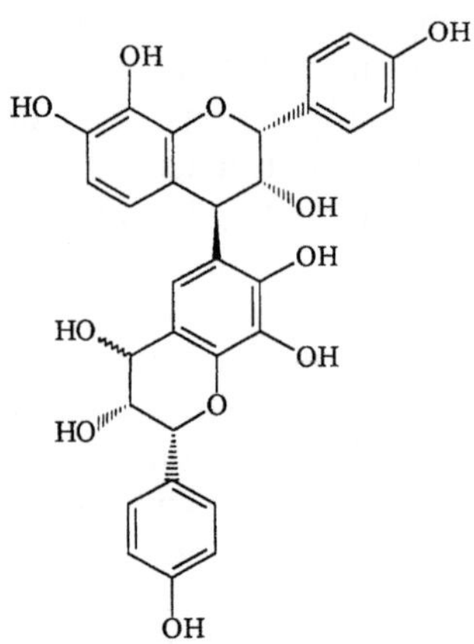

(12)
(13)

The elegance of this simple biomimetic approach to the synthesis of proanthocyanidin oligomers was demonstrated during synthesis of the 'mixed' profisetinidin trimers (**22**) and (**23**), *i.e.* analogues possessing different ABC and GHI chain extender units. Triflavanoid (**22**) with its fisetinidol ABC and epifisetinidol GHI units was formed by acid-catalyzed reaction of fisetinidol-(4α→8)-catechin (**24**) (*51*) and epifise-

References, pp. 59–67

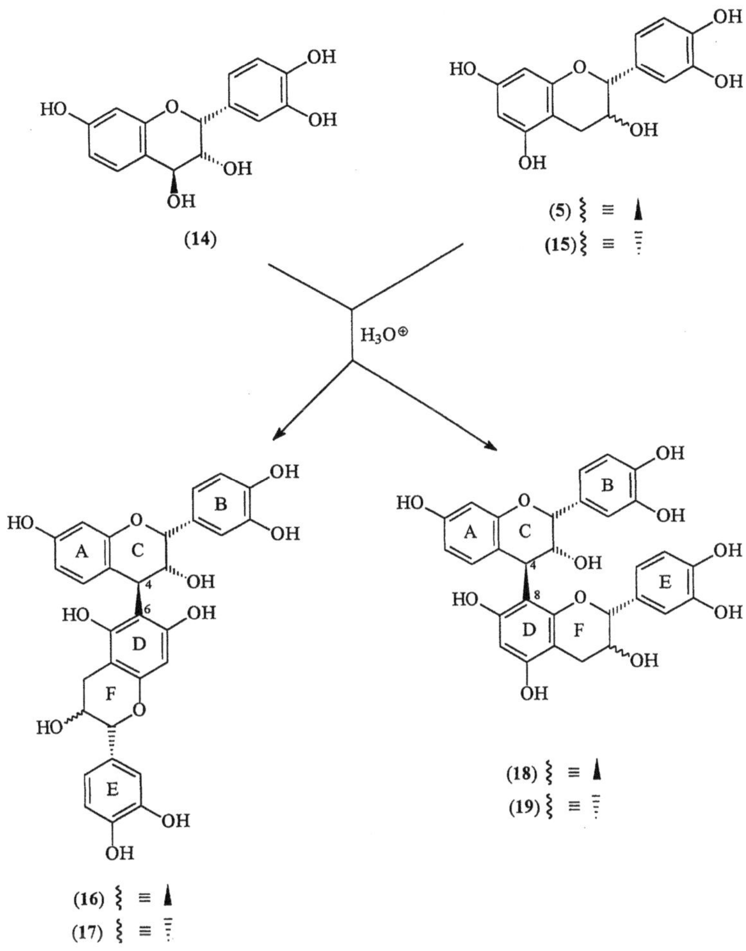

Scheme 2. Synthesis of dimeric profisetinidins (**16**)–(**19**) with epifisetinidol chain extender units

tinidol-4β-ol (**14**). The remaining triflavanoid (**23**) with its epicatechin DEF unit was similarly synthesized using epifisetinidol-(4β→6)-epicatechin (**17**) (*vide supra*) in the acid-catalyzed condensation with fisetinidol-4α-ol (**25**). Comparison of the ^{1}H NMR and CD data of the permethylaryl ether triacetates of (**22**) and (**23**) with those of the same derivatives of the natural products again provided unequivocal structural proof for the latter compounds. It should be emphasized that interpretation of CD data, *i.e.* the sign and amplitude of the Cotton effect in the 220–250 nm region, beyond the dimeric level should be restricted to comparative applications.

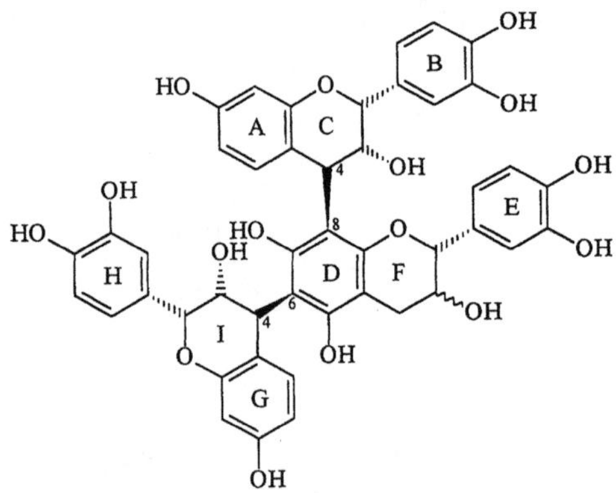

(20) ⌇ ≡ ▲

(21) ⌇ ≡ ⋮

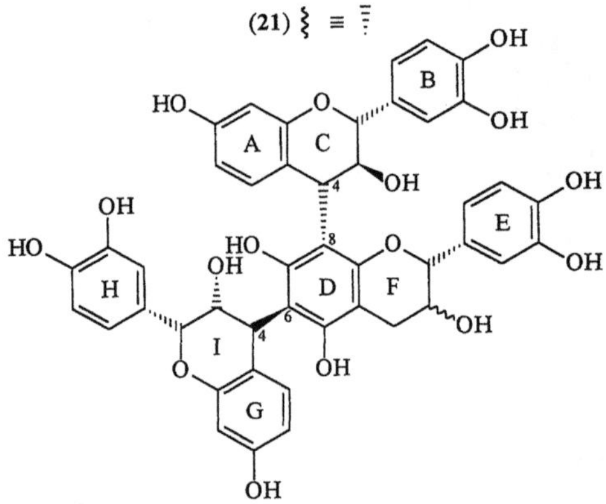

(22) ⌇ ≡ ▲

(23) ⌇ ≡ ⋮

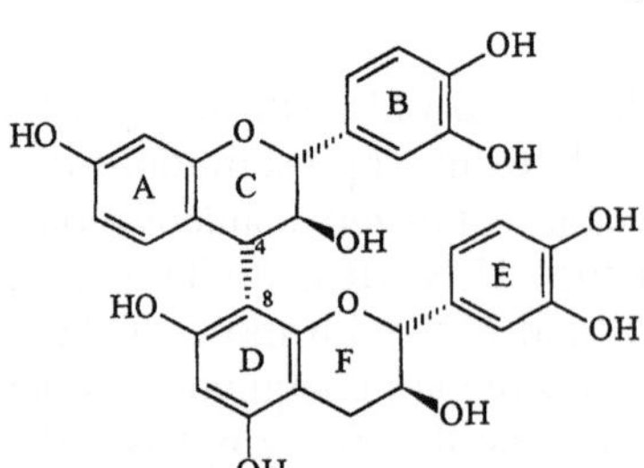

(24)

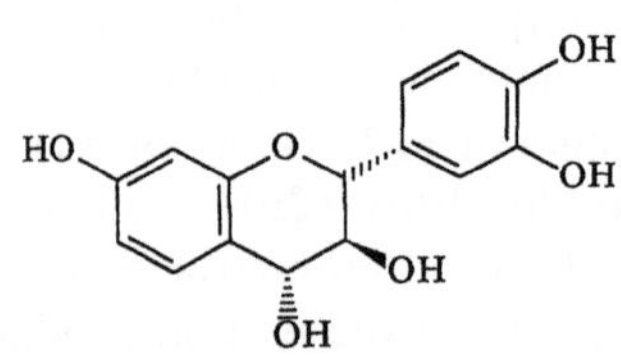

(25)

2.2. Synthesis of Procyanidins under Neutral Conditions

Owing to the lability of the interflavanyl bond in procyanidins under either acidic or basic conditions, the existing semi-synthetic methods (*28, 29*) invariably result in an equilibrium between substrates and products. Such a labile bond and the apparent preference of the electrophile for the di- and trimeric products (see Scheme 1), once condensation is initiated, furthermore give poor control over the level of oligomerization. We thus assessed (*52*) the effectiveness of the thiophilic Lewis acids, dimethyl(methylthio)sulfonium tetrafluoroborate (DMTSF) (*53, 54*) and silver tetrafluoroborate ($AgBF_4$)(*55*) to activate the C(4)-S bond in the 4-thioethers of flavan-3-ols towards carbon nucleophiles and hence to generate the interflavanyl bond of procyanidins under neutral conditions.

Treatment of a mixture of 4β-benzylsulfanylepicatechin (**26**), representing the (2*R*,3*R*)-2,3-*cis*-flavan-3-ol chain extender unit of the procyanidins, and catechin (**5**) in THF with DMTSF or $AgBF_4$ at or below 0°C afforded procyanidin B-1 (**27**) (22 and 38% for DMTSF and $AgBF_4$ respectively) and the analogous trimeric procyanidin (**28**) (10% for DMTSF, not formed with $AgBF_4$) (Scheme 3). The moisture sensitivity of DMTSF and the improved yield of (**27**) recorded for $AgBF_4$ prompted the use of the latter Lewis acid in further work. When a mixture of 4β-benzylsulfanylepicatechin (**26**) and epicatechin (**15**) in THF was treated with $AgBF_4$, procyanidin B-2 (**29**) was formed in 37% yield without evidence of the formation of regioisomeric dimers or of higher oligomers. This protocol thus compares favourably with the classical acid-catalyzed condensation of catechin-4α-ol and catechin (**5**) (*40, 56*) which gave a mixture of procyanidin B-3 (**32**) and B-6, the trimeric procyanidin C-2 (**33**) (*cf.* Scheme 4) and its 4,6-regioisomer, and the presumed all-*trans*-(4→8)-linked tetraflavanoid analogue (10:1:12:1:3) (45% overall yield).

The scope of the thiophilic Lewis acid mediated interflavanyl bond formation was extended using 4-benzylsulfanylcatechin (**30**) (4:1 mixture of 4β- and 4β-epimers) (*40, 56*), representing the (2*R*,3*S*)-2,3-*trans*-flavan-3-ol chain extender unit of the procyanidins, as source of the flavan-3-ol C(4) electrophilic moiety (Scheme 4). Separate treatment of a mixture of the epimeric 4-benzylsulfanylcatechins (**30**) and catechin (**5**) and epicatechin (**15**) in THF with $AgBF_4$ afforded procyanidin B-3 (**32**) (35%) and B-4 (**31**) (51%), respectively. The preference for the formation of 4β- and 4α-interflavanyl bonds using the epicatechin- and catechin-4-thiobenzyl ethers (**26**) and (**30**), respectively, and for the (4→8)-interflavanyl linkages were anticipated (*40, 56*), *i.e.* the thio-

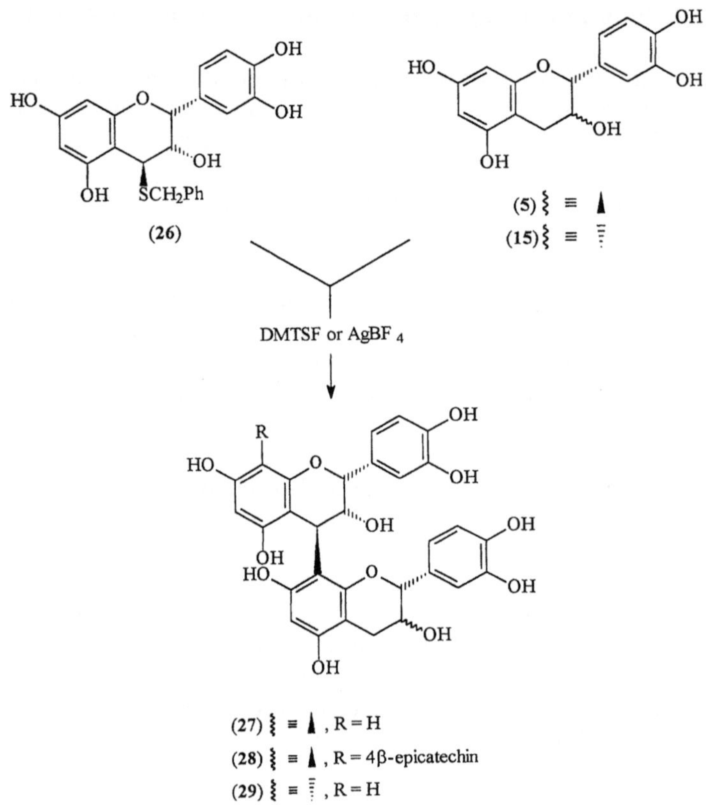

Scheme 3. Synthesis of procyanidin B-1 (**27**) and B-2 (**29**) using Lewis acid activation of 4β-benzylsulfanylepicatechin (**26**)

benzyl ethers are converted by the Lewis acids into relatively stable intermediates permitting both regioselective attack of the nucleophile *via* C(8) where the HOMO displays maximum amplitude, and stereoselectivity by approach from the sterically least hindered side.

Utilization of procyanidin B-3 (**32**) as nucleophile in a coupling reaction with the 4-benzylsulfanylcatechin epimers (**30**) using $AgBF_4$ in THF, gave the trimeric procyanidin (**33**) (26%) as the only isolable product (Scheme 4).

The sequence towards the procyanidins depicted in Schemes 3 and 4 using $AgBF_4$ as the thiophilic Lewis acid no doubt offers advantages as far as control over the level of oligomerization, reversibility and 'scattering' of the interflavanyl bond(s) are concerned in comparison

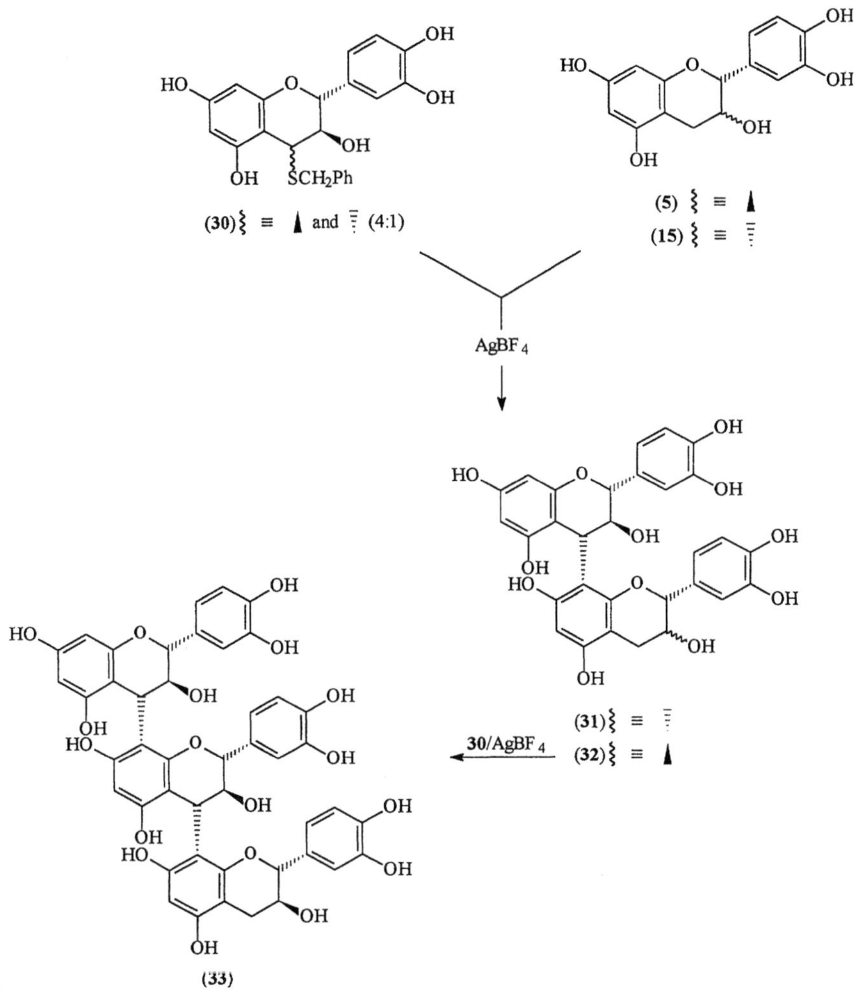

Scheme 4. Synthesis of procyanidin B-3 (**32**), B-4 (**31**) and C-2 (**33**) using Lewis acid activation of the 4-benzylsulfanylcatechin epimers (**30**)

with the formation of these products under conditions previously developed [(*28*), and references cited therein).

2.3. Synthesis of Ether-Linked Proteracacinidins

Proanthocyanidins possessing ether-type interflavanyl linkages are extremely rare except for the A-type oligomers which contain the conventional C(4)–C(6)/(8) bond as well as an additional ether linkage connecting C(2) (C-ring) and C(5)/C(7) (D-ring) (see Sections 2.5 and 3.2). Analogues possessing exclusively ether bonds are so far restricted to the 1,4-dioxane type dimeric profisetinidins (**34**) and (**35**) (*57, 58*), the (C_4-O-C_4)-promelacacinidins (**36**) and (**37**) (*59*), and the (4→7:5→6) doubly-linked proteracacinidin-type analogue (**38**) (*60*). We recently identified the first two [4-O-3]-linked bis-teracacinidins (**39**) and (**40**) (*61*) and two related [4-O-4]-linked analogues (**41**) and (**42**) (*62*) and subsequently explored possible semi-synthetic routes in order to establish their stereochemistry.

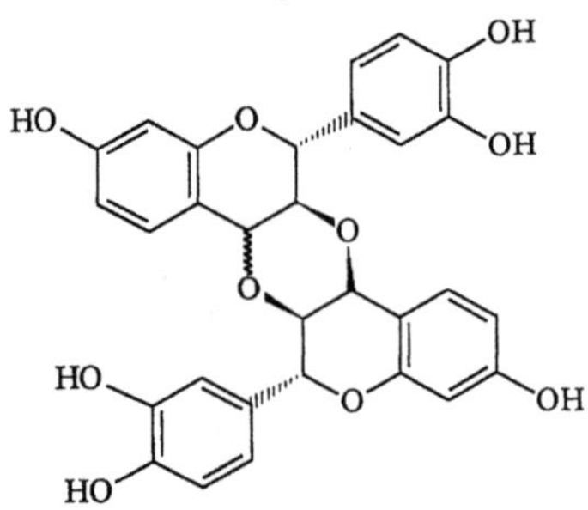

(**34**)
(**35**)

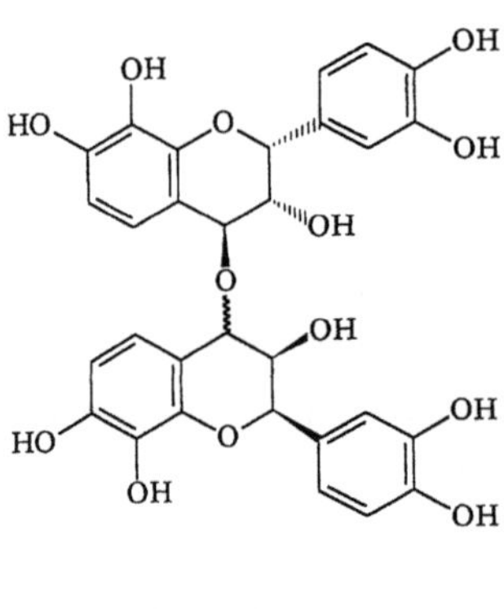

(**36**)
(**37**)

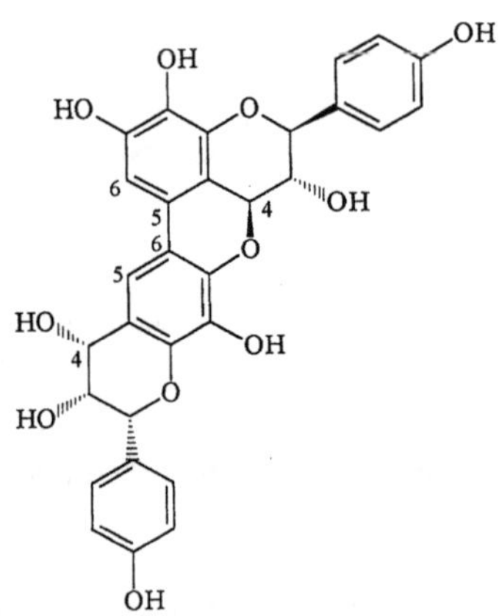

(**38**)

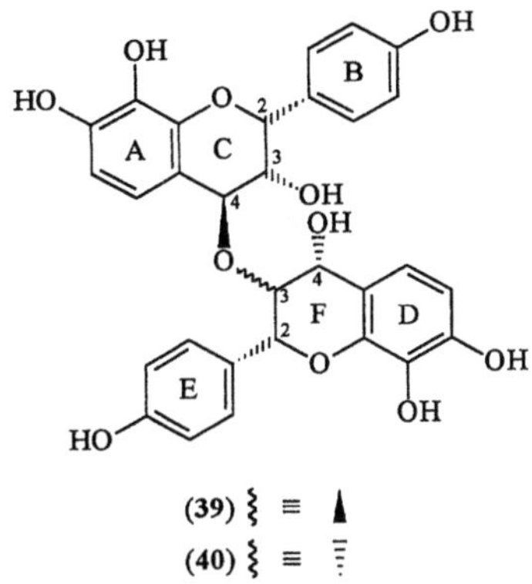

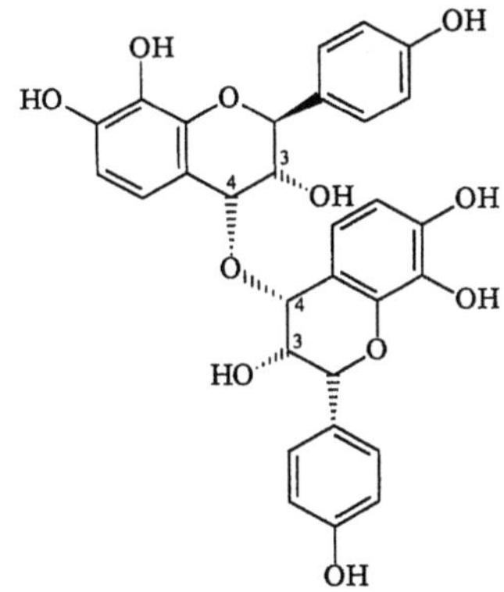

(41)

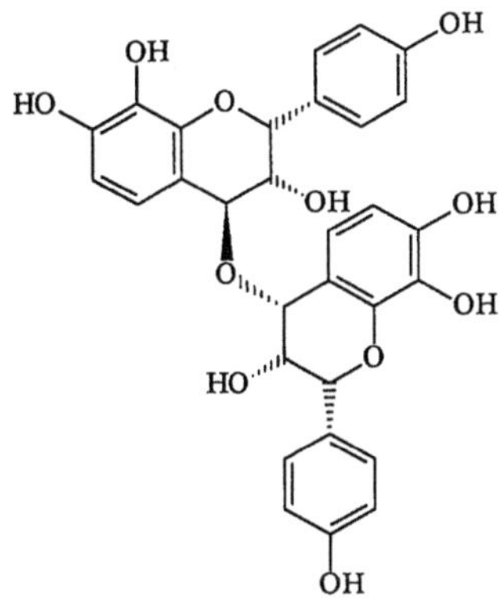

(42)

It was anticipated that the C(4) benzylic ether bonds in proteracacinidins (**39**)–(**42**) would be susceptible to solvolysis in aqueous medium. The very same conditions which are applied universally for the formation of $C(sp^3) \rightarrow C(sp^2)$ interflavanyl linkages (*40*), would hence be less applicable to the generation of the ether bonds in compounds (**39**)–(**42**). We thus opted to enhance the electrophilicity at C(4) of one of the flavan-3,4-diol methyl ethers, *e.g.* (**43**) by formation of the 4-chloroflavan-3-ol derivative (**44**) in order to permit the formation of the crucial ether bond at near a neutral pH value.

The synthesis of the [4-O-3]-linked proteracacinidin derivative (**46**) is delineated in Scheme 5. Epioritin-4α-ol tri-*O*-methyl ether (**43**) with

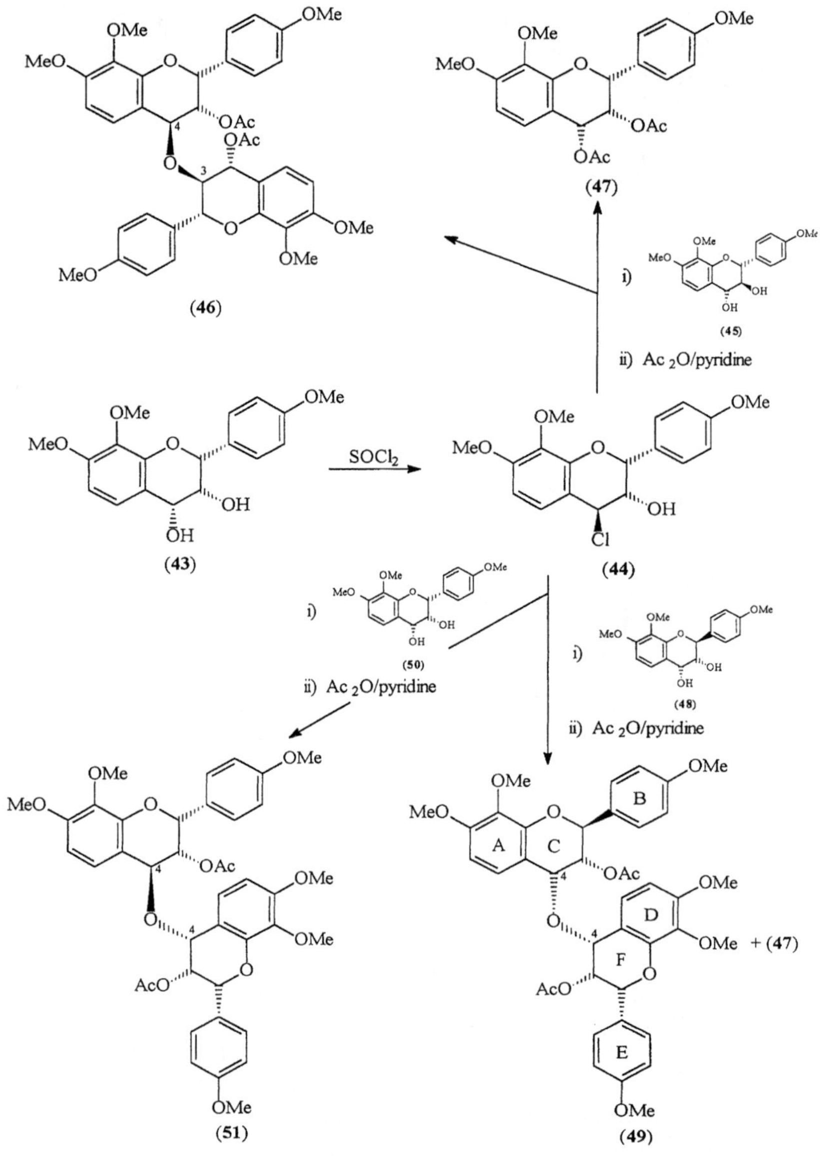

Scheme 5. Synthesis of the ether-linked proteracacinidin derivatives (**46**), (**49**) and (**51**)

2*R*,3*R*,4*R* absolute configuration was converted quantitatively into the intermediate (2*R*,3*S*,4*S*)-4-chloroflavan-3-ol (**44**) with thionyl chloride in dry THF, with the anticipated inversion of configuration at C(4) of precursor (**43**). Addition of a two molar excess of oritin-4α-ol tri-*O*-methyl ether (**45**) and eventual acetylation gave the all-*cis* flavan-3,4-diacetate (**47**) and the epioritin-(4β→3)-oritin-4α-ol permethylaryl ether diacetate (**46**) (15%) with ^{1}H NMR and CD data identical with those of the same derivative of the natural product (**39**). Separate treatment of the intermediate 4β-chloroflavan-3-ol derivative (**44**) with *ent*-oritin-4α-ol tri-*O*-methyl ether (**48**) and epioritin-4α-ol tri-*O*-methyl ether (**50**) in dry THF and subsequent acetylation afforded *ent*-oritin-(4α→4)-epioritin-4α-ol permethylaryl ether diacetate (**49**) (15%) and the epioritin-(4β→4)-epioritin-4α-ol permethylaryl ether diacetate (**51**) (24%), respectively. Compounds (**49**) and (**51**) were identical with the same derivatives of the natural products (**41**) and (**42**), respectively, by comparison of their ^{1}H NMR and CD data.

The stereoselective coupling between the 4-chloroflavan-3-ol derivative (**44**) and the flavan-3,4-diol derivatives (**45**) and (**50**) to give the C(4)-O-C(3)- and C(4)-O-C(4)-proteracacinidins (**46**) and (**51**), respectively, with retention of the C(4) configuration of the electrophilic precursor (**44**), is explicable in terms of a neighboring group mechanism involving intramolecular displacement of the *quasi*-axial C(4)-chloro nucleofuge by the axial C(3)-hydroxyl group. The transient protonated epoxide (**52**) then permits preferential attack of the nucleophilic C(4)-hydroxyl group of the flavan-3,4-diol derivatives (**45**) and (**50**) from the less hindered β-face resulting in a highly stereoselective coupling step.

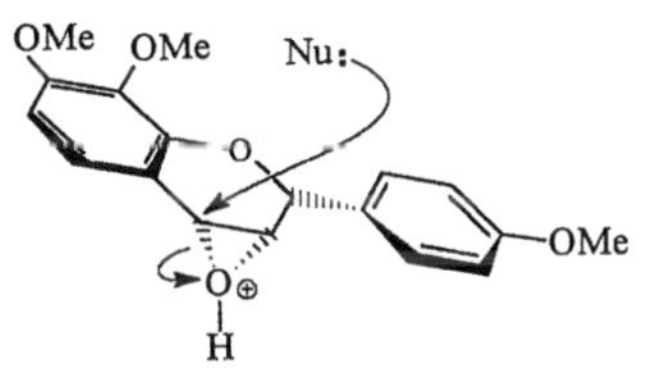

(52)

It was anticipated that coupling of the 4β-chloroflavan-3-ol (**44**) and the *ent*-oritin-4α-ol derivative (**48**) would also proceed *via* the neighboring group mechanism. The rather unexpected formation of the 4α-ether bond (F-ring) in (**49**), *i.e.* with inversion of configuration at C(4) of the electrophile (**44**), presumably reflects reaction conditions incapable of triggering the neighboring group mechanism, hence resulting in an S_N2-type mechanism which requires the approaching hydroxyl nucleophile to force out the nucleofugal chloride. The requisite alignment for such a concerted process may be facilitated by mutual hydrogen bonding of the C(3)-hydroxyl groups of (**44**) and (**48**) which is effectively permitted by the *axial* C(4)-hydroxyl group of flavan-3,4-diol (**48**) compared to the *equatorial* orientation of the same functionality in nucleophiles (**45**) and (**50**). In addition, unfavorable 1,3-*diaxial* interaction between $H\text{-}2_{ax}$ of a putative oxirane of type (**52**) and the approaching nucleophile (**48**) would not favor a neighboring group mechanism.

The formation of the all-*cis*-flavan-3,4-diol derivative, isolated as diacetate (**47**), in the coupling reactions of the 4β-chloroflavan-3-ol (**44**) and diol derivatives (**45**) and (**48**) is explicable in terms of solvolysis of the remaining electrophile (**44**) during work-up and chromatography. Inversion of configuration is effected by intermolecular hydrogen bonding between the *axial* C(3)-hydroxyl group and water hence permitting S_N2 displacement of chloride ion.

The co-occurrence of the ether-linked proteracacinidins (**39**)–(**42**) as well as the 'conventional' carbon coupled analogues in *Acacia galpinii* (*44*) presumably reflects the poor nucleophilicity of the pyrogallol A-ring of the monomeric flavan-3,4-diol precursors thus permitting alternative centres to participate in interflavanyl bond forming processes.

2.4. Miscellaneous

The principle of condensing electrophilic and nucleophilic flavanyl units under mild acidic conditions earlier culminated in a unique series of papers (*63–67*) dealing with the synthesis of a range of profisetinidin-type tetraflavanoids, *e.g.* (**53**) and (**54**). Conformational analysis of the permethylaryl ether tetra-acetates of tetraflavanoids (**53**) and (**54**) as well as those of related analogues revealed overall 'cyclic' arrangements of flavanyl units in each despite mutual stereochemical and structural differences (*66, 67*). The unique thermodynamic stability of their dominant conformers was attributed to the combined effects of the relative configurations of constituent flavanyl units, to steric repulsion by functional groups *ortho* to interflavanyl bonds, and to steric inhibition of mobility about interflavanyl bonds due to partial overlap of terminal units.

References, pp. 59–67

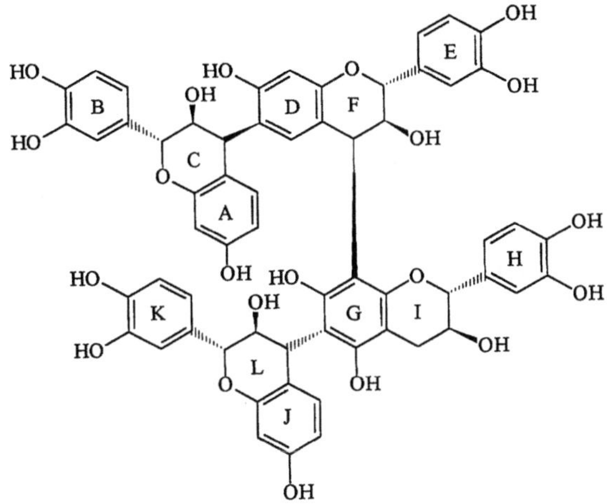

(53)

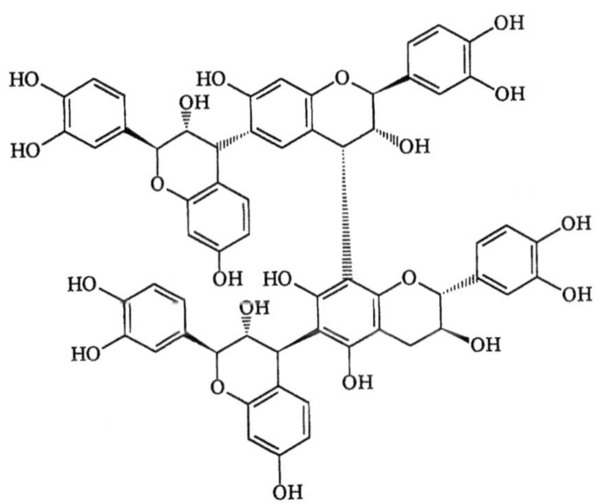

(54)

A similar synthetic approach was also implemented to synthesize a range of related proanthocyanidins. Notable among these are the synthesis of dimers, *e.g.* (**55**), exhibiting flavan chain extender units derived from flavan-4-ols, *e.g.* (**56**), as electrophiles (*68*, *69*), 'conventional' proguibourtinidin-type biflavanoids, *e.g.* the guibourtinidol-(4α→8)- and (4β→8)-afzelechins (**57**) and (**58**) (*70*), and procyanidins ^{2}H-labeled at C(4) (C-ring), *e.g.* (**59**), using catechin-4-ols with C(4)-deuterium labels which are available by reduction of (+)-taxifolin with sodium borotetradeuteride (*71*). In the latter instance formation of the interflavanyl bond between the perbenzylaryl ethers was catalyzed by the Lewis acid titanium tetrachloride in dichloromethane. The free phenolic deuterium labeled procyanidin B-3 (**59**) was then generated by catalytic hydrogenolysis of the benzyl protecting groups.

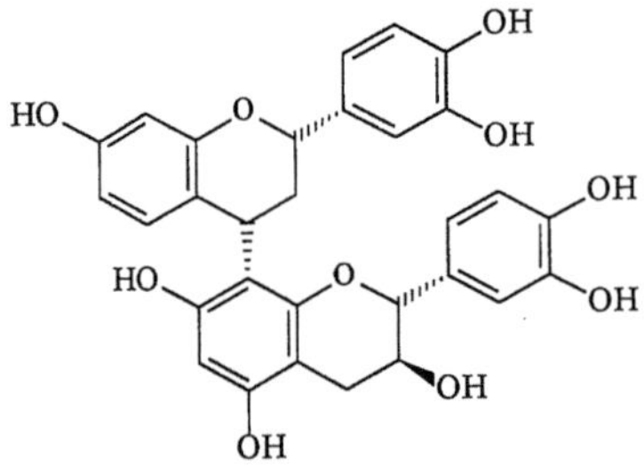

(**55**)

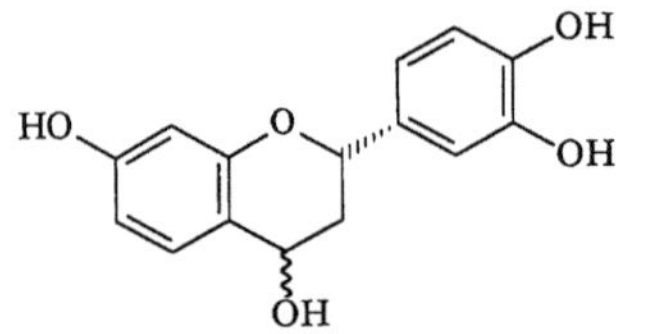

(**56**)

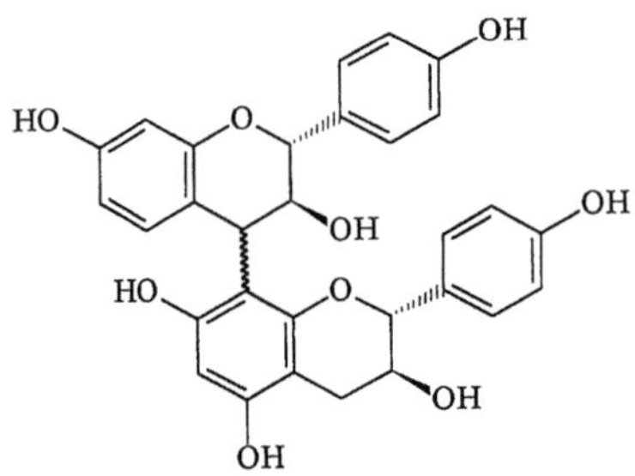

(**57**)

(**58**)

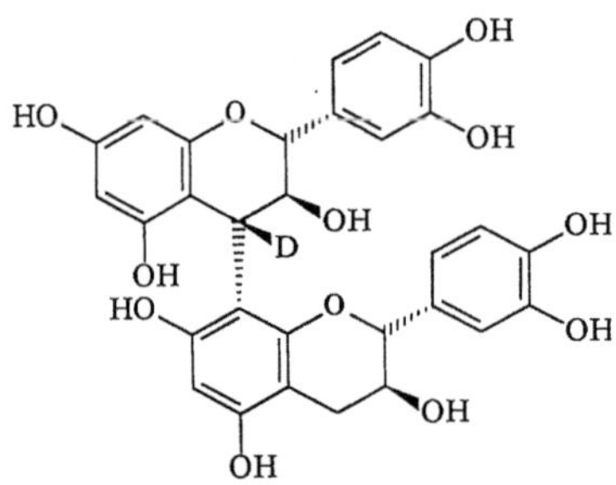

59

Another notable achievement is the synthesis of a high molecular mass condensed tannin that is based upon the cationic polymerization of 3′,4′,5,7-tetrabenzyloxyflavan-3,4-carbonate (**60**), available in 92% yield *via* treatment of the parent flavan-3,4-diol with N,N′-carbonyldiimidazole, using Lewis acid (*e.g.* BF_3, $TiCl_4$, $SbCl_5$, $Ph_3C \cdot BF_4$)

catalysis (*72*). Debenzylation was accomplished by catalytic hydrogenation using 10% Pd/C in 20% ethanol-THF at room temperature. The number-average molecular mass (Mn) of the acetylated synthetic condensed tannin was determined as 3794–17367 by gel permeation chromatography (GPC), using a polystyrene standard. Thus, the molecular mass of the free phenolic form was estimated to be 2209–10200.

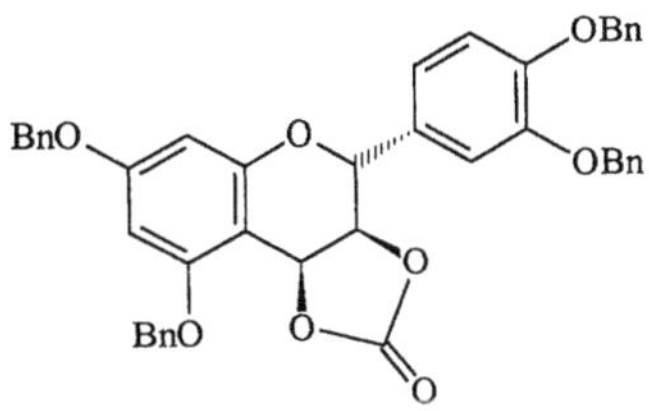

(60)

2.5. Formation of the Ether Linkage in A-Type Proanthocyanidins

In contrast to the aforementioned proanthocyanidins of the B-type where the constituent flavanyl units are linked by only one bond, analogues of the A-class possess an unusual second ether linkage to C(2) of the T-unit. Studies focussing on the oxidative formation of this bond are surprisingly limited and are hitherto restricted to the use of hydrogen peroxide (*73*) and dioxygen (*74–76*) as oxidants.

The structures of proanthocyanidin A-6 (**62**) [epicatechin-(2β→7:4β→6)-epicatechin] and A-7 (**63**) [epicatechin-(2β→5:4β→6)-epicatechin] [see ref. (*77*) for nomenclature] were unequivocally confirmed by oxidative conversion of procyanidin B-5 (**61**) [epicatechin-(4β→6)-epicatechin] using hydrogen peroxide in aqueous sodium hydrogen carbonate solution (*73*).

Similar treatment (*74*, *76*) of robinetinidol-(4β→8)-catechin mono-*O*-methyl ether (**64**) in basic reaction medium (pH 10.0) but with dioxygen as the oxidant led to the formation of robinetinidol-(2β→7:4β→8)-catechin mono-*O*-methyl ether (**66**) (18% yield), the first A-type analogue of the 5-deoxy (A-ring) series of oligoflavanoids, and also the first entry amongst this class of proanthocyanidins with a 3,4-*cis* C-ring configuration. Comparison of the ^{1}H NMR data of compound (**66**) with those of other A-type proanthocyanidins with 3,4-*trans* C-ring orientations revealed the conspicuous identity of their 3- and 4-H vicinal coupling constants ($^3J_{3,4}$ 3–4 Hz). This phenomenon is explicable in terms of the conformational rigidity of the bicyclic ring

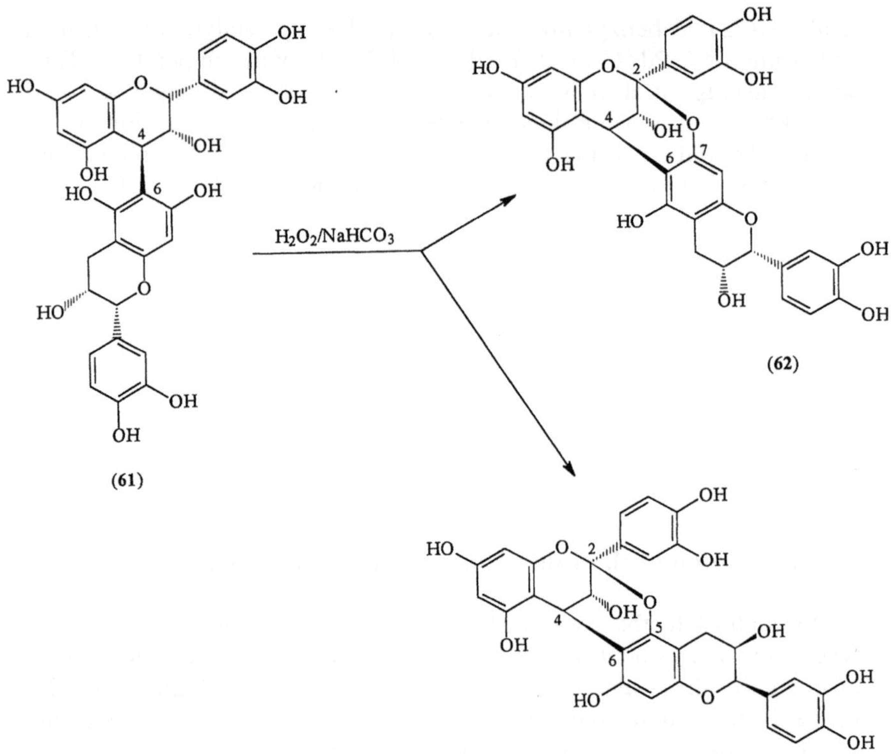

(61)

(62)

(63)

system which culminates in very similar dihedral angles between 3- and 4-H(C) in both 3,4-*trans* and 3,4-*cis* homologues which should thus lead to almost identical coupling constants for these protons. The C(4)-O (E-ring) demethyl analogue of compound (**66**) was eventually identified in commercial wattle bark extract (*76*).

The transformation of the B-type prorobinetinidin (**64**) into the A-type analogue (**66**), presumably involves oxidative removal of hydride ion at C(2) (C-ring) as the initial step. The intermediate B-ring quinone methide (**65**) is then susceptible to cyclization involving 7-OH(D) and the *si*-face at C(2), hence generating the second interflavanyl ether-type linkage. The nature of the oxidizing species is unclear. Although dioxygen in itself may effect the transformation, (**64**)→(**65**), the prevailing conditions may alternatively induce oxidation of the *o*-dihydroxyfunctionality of the pyrogallol B-ring to an *o*-quinone (*78*) which subsequently serves as oxidant for the conversion, (**64**)→(**65**). A

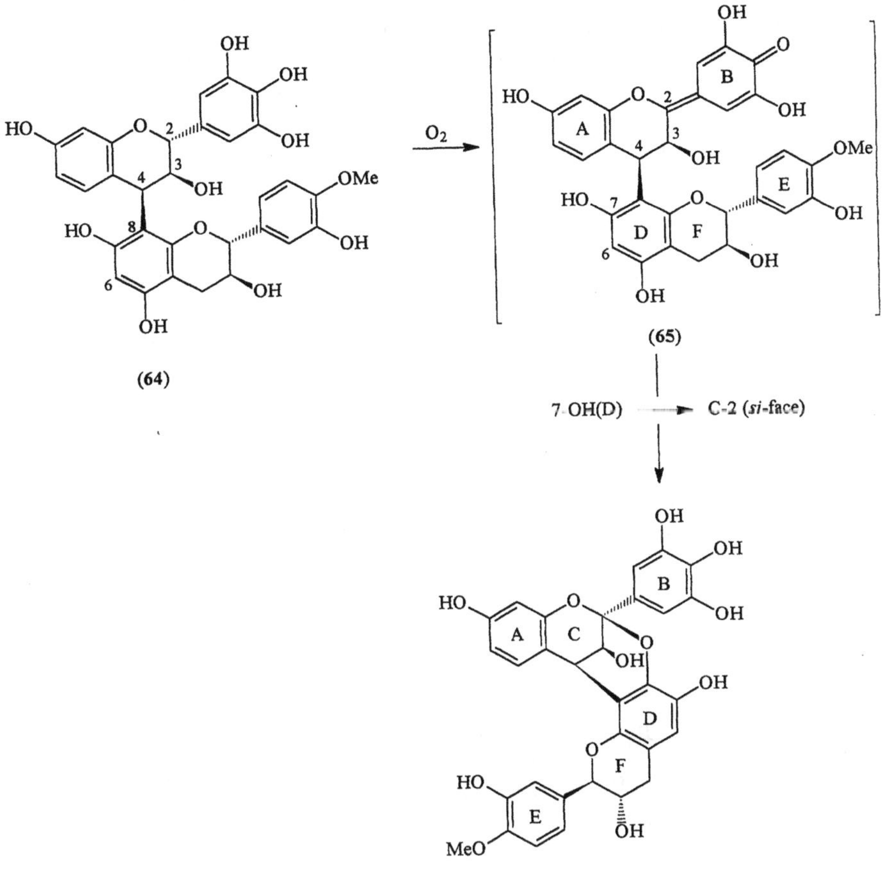

(64) (65) (66)

similar B- to A-type conversion was also observed when procyanidin B-2 was subject to dioxygen in mild basic reaction medium (*75*).

3. Cleavage of the Interflavanyl Bond in Oligomeric Proanthocyanidins

3.1. B-Type Proanthocyanidins

The acid-catalyzed cleavage of the interflavanyl bond in proanthocyanidins exhibiting C(5)-oxygenation of the A-ring of their chain-extender units with sulfur (*79*, *80*) and oxygen (*81*) capturing nucleophiles and yielding flavan-3-ol 4-thioethers (by use of sulfur nucleophiles) from the extender units and flavan-3-ols from the terminal unit, has played a crucial role in structure elucidation of these complex natural products. In the 5-deoxy (A-ring) series of compounds, *e.g.* the fisetinidol-(4→8)- and (4→6)-catechin profisetinidins (**24**), (**69**), and (**71**), and the analogous prorobinetinidins (**67** and **72**) from the commercially important bark of *Acacia mearnsii* (black wattle) (*30, 31*), this $C(sp^3)$–$C(sp^2)$ bond is remarkably stable under a variety of conditions (*51*) and has hitherto resisted all efforts at cleavage in a controllable manner. Such a stable interflavanyl bond hampered both the structure investigation of the polyflavanoid tannins in black wattle bark and those from other commercial sources, *e.g. Schinopsis* spp. (quebracho), as well as the establishment of the absolute configuration of the chain-terminating flavan-3-ol moiety in the 5-deoxyoligoflavanoids. We therefore assessed conditions to efficiently cleave the interflavanyl bond in profisetinidins under conditions sufficiently mild to allow the isolation and identification of the constituent flavanyl units (*82, 83*).

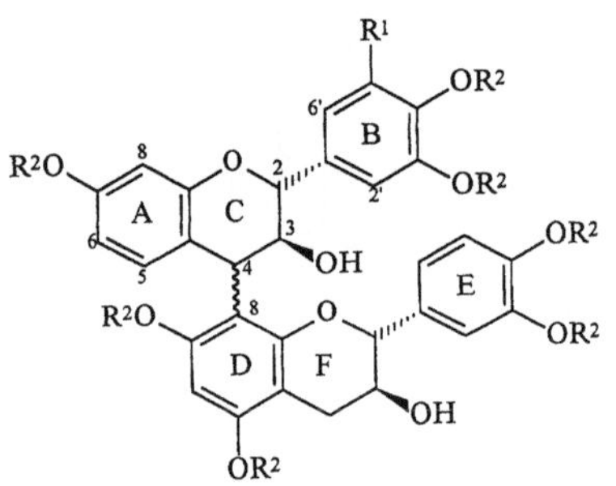

(**24**) ≀ ≡ ⋮ , $R^1 = R^2 = H$
(**67**) ≀ ≡ ⋮ , $R^1 = OH, R^2 = H$
(**68**) ≀ ≡ ⋮ , $R^1 = H, R^2 = Me$
(**69**) ≀ ≡ ▲ , $R^1 = R^2 = H$
(**70**) ≀ ≡ ▲ , $R^1 = H, R^2 = Me$

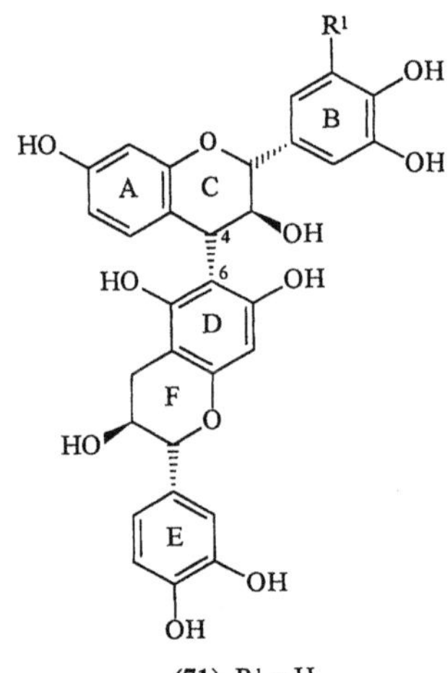

(**71**) $R^1 = H$
(**72**) $R^1 = OH$

Treatment of the fisetinidol-(4α→8)-catechin (**24**) (*51*), representing a typical tannin unit of commercial wattle extract, with sodium cyanoborohydride [$Na(CN)BH_3$] (*84*) in trifluoroacetic acid (TFA) for 6 h at 0°C gave products comprising the starting material (**24**), catechin (**5**) (15%) and the (2*R*)-1-(2,4-dihydroxyphenyl)-3-(3,4-dihydroxyphenyl)propan-2-ol (**77**) (Scheme 6). Similar treatment of the fisetinidol-(4β→8)- and -(4α→6)-catechins (**69**) and (**71**) (*51*) with their respective

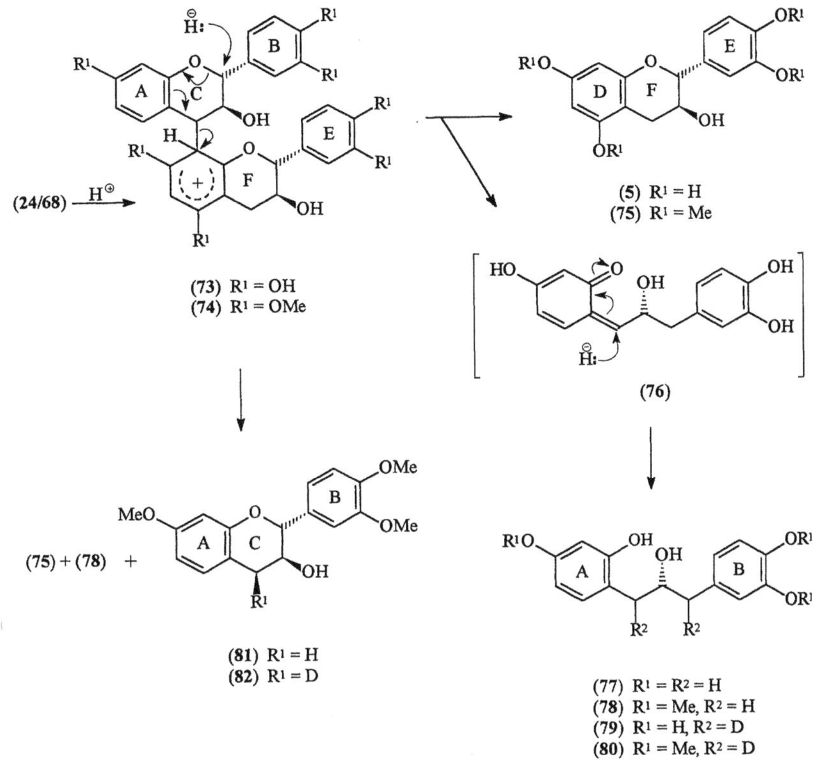

Scheme 6. Proposed route to the cleavage of the interflavanyl bond and of the C-ring in profisetinidins *e.g.* (**24**) and permethylaryl ether (**68**)

more and less labile interflavanyl bonds compared with the C(4)–C(8) bond in compound (**24**) under acidic conditions (*85*) also afforded a mixture consisting of starting material (**69**) and (**71**), catechin (**5**) (17, 4% resp.), and the (2*R*)-1,3-diarylpropan-2-ol (**77**) (18, 4%) resp.).

Similar conditions also effected cleavage of the interflavanyl bond in the fisetinidol-(4α→8)-catechin permethylaryl ether (**68**) to afford tetra-*O*-methylcatechin (**75**) (21%), the 1,3-diarylpropan-2-ol (**78**) (12%), and tri-*O*-methylfisetinidol (**81**) (12%). Such a rupture of the interflavanyl bond in the permethylaryl ether (**68**) introduced an important dimension to these cleavages in relation to the chemistry of the 5-deoxyoligoflavanoids where the additional chromatographic steps involved with derivatization are often prerequisites for sample purity. The 'liberation' of the chain-terminating flavan-3-ol unit (**5**) or (**75**), irrespective of whether the phenol (**24**) or methyl ether (**68**) was employed, provides a powerful probe towards addressing the hitherto unsolved problem of defining the absolute configuration at the stereocenters of this moiety in naturally occurring proanthocyanidins that are synthetically inaccessible.

The mild conditions which thus effect simple cleavage of the strong interflavanyl bond in the profisetinidins (**24**), (**69**) and (**71**) prompted application of the same protocol to the procyanidins B-1 (**27**) and B-3 (**32**) and their respective permethylaryl ethers (**83**) and (**84**) with less rigid C(4)–C(8) linkages compared to those in the profisetinidins (**24**) and (**69**). Treatment of procyanidin B-1 (**27**) with $Na(CN)BH_3$ in TFA for 1 h at 0°C gave a mixture comprising the starting material (**27**), catechin (**5**) (20%), and epicatechin (**15**) (21%). Under identical conditions, procyanidin B-3 (**32**) afforded catechin (**5**) (35%) and a residue of starting material. The permethylaryl ethers (**83**) and (**84**) gave, within 30 min, respectively tetra-*O*-methylcatechin (**75**) (31%), tetra-*O*-methylepicatechin [C(3)-epimer of (**75**)] (33%), and starting material (**83**), and tetra-*O*-methylcatechin (**75**) (56%) and starting material (**84**).

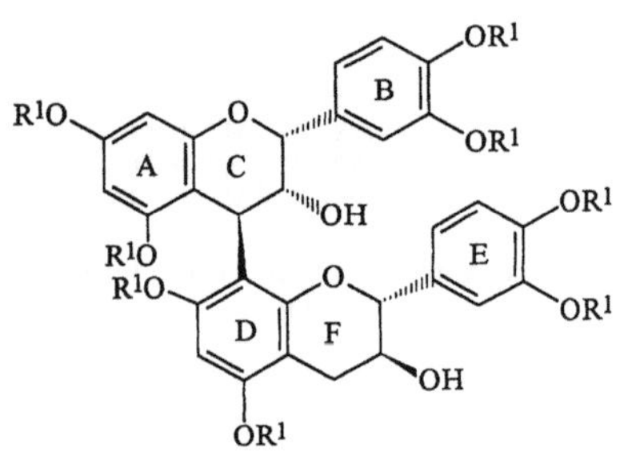

(**27**) R1=H
(**83**) R1=Me

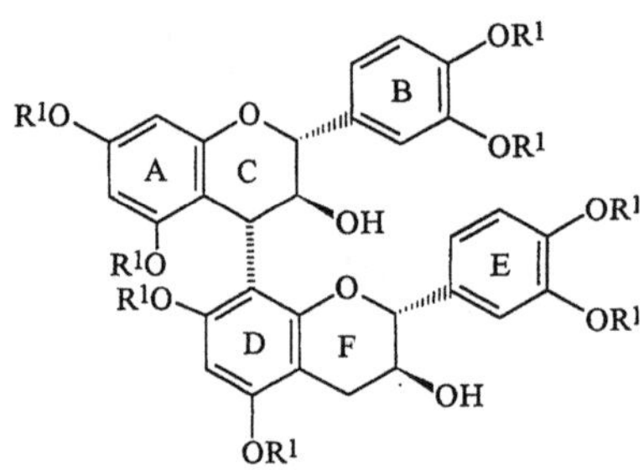

(**32**) R1=H
(**84**) R1=Me

Whereas the heterocyclic ring of the catechin DEF moiety invariably remains intact during the reductive process, cleavage of both (4→6)- and (4→8)-interflavanyl bonds in the free phenolic profisetinidins (**24**), (**69**), and (**71**) is apparently associated with simultaneous opening of the C-ring of the chain-extender unit. Protonation of the electron-rich phloroglucinol D-ring (*86, 87*) in profisetinidin (**24**) (Scheme 6), and concomitant delivery of the equivalent of a hydride ion at C(2) (C-ring) of intermediate (**73**) effects the concurrent rupture of the pyran C-ring and of the C(4)–C(8) bond to give catechin (**5**) and the *o*-quinone methide intermediate (**76**), which is subsequently reduced to the 1,3-diarylpropan-2-ol (**77**). The selective cleavage of the interflavanyl bonds in procyanidins B-1 (**27**) and B-3 (**32**), and their permethylaryl ethers (**83**) and (**84**) presumably results from the relative lability of this bond, imposing a high degree of S_N1 character to the processes of protonation and delivery of hydride ion.

The mechanism for cleavage of the interflavanyl bond in the profisetinidin biflavanoids (Scheme 6) was corroborated using sodium cyanotrideuterioborohydride [$Na(CN)BD_3$] in TFA. Under these conditions the fisetinidol-(4α→8)-catechin (**24**) was converted into catechin (**5**) (26%) and the (2*R*)-1,3-dideuterio-1,3-diarylpropan-2-ol (**79**) (25%), while the permethylaryl ether (**68**) and the fisetinidol-(4β→8)-catechin hepta-*O*-methyl ether (**70**) both gave tetra-*O*-methylcatechin (**75**) (12, 32% resp.), the dideuterio-1,3-diarylpropan-2-ol tri-*O*-methyl ether (**80**) (14, 16% resp.) and the 4β-deuteriofisetinidol derivative (**82**) (12, 14% resp.). Formation of the deuteriated, 1,3-diarylpropan-2-ols (**79**) and (**80**) (mixtures of diastereomers) thus confirmed the conjecture regarding the genesis of the propan-2-ols *via* reduction of the *o*-quinone methide (**76**).

The protonated species (**73/74**) presumably also served as precursor to the 4β-deuteriotri-*O*-methylfisetinidol (**82**) by delivery of hydride ion from the β-face in a predominant S_N2 mode. Compound (**82**) persistently formed also when fisetinidol-(4α→8)- and (4β→8)-catechin hepta-*O*-methyl ethers (**68**) and (**70**) were treated with $Na(CN)BD_3$ in TFA. This observation prompted an investigation of the structural features of the substrates that direct the stereochemistry of the delivery of hydride ion at C(4) in intermediates of type (**73/74**). Whereas treatment of the epifisetinidol-(4β→8)-catechin hepta-*O*-methyl ether (**85**) with $Na(CN)BD_3$ afforded the 4β-deuteriotri-*O*-methylepifisetinidol (**87**) (18.5%), tetra-*O*-methylcatechin (**75**) (32%) and the (2*S*)-1,3-dideuterio-1,3-diarylpropan-2-ol [6%, enantiomer of compound (**80**)], the *ent*-fisetinidol-(4β→8)-catechin hepta-*O*-methyl ether (**86**) gave 4α-deuteriotri-*O*-methyl-*ent*-fisetinidol [13%, the enantiomer of compound (**82**)], tetra-*O*-methylcatechin (**75**) (24%) and the (2*S*)-1,3-dideuterio-1,3-diarylpropan-2-ol [12%, enantiomer of (**80**)].

Thus, the formation of the 4β-deuteriofisetinidol- and epifisetinidol derivatives (**82**) and (**87**) from the reduction of the profisetinidin permethylaryl ethers (**68**), (**70**), and (**85**) with $Na(CN)BD_3$ in TFA, and of the enantiomer of compound (**82**) during reduction of the *ent*-fisetinidol-(4β→8)-catechin derivative (**86**), indicated that the deuterium ion is consistently delivered to C(4) of a protonated species of type (**73/74**) from the side opposite to the 2-aryl group of the C-ring. This

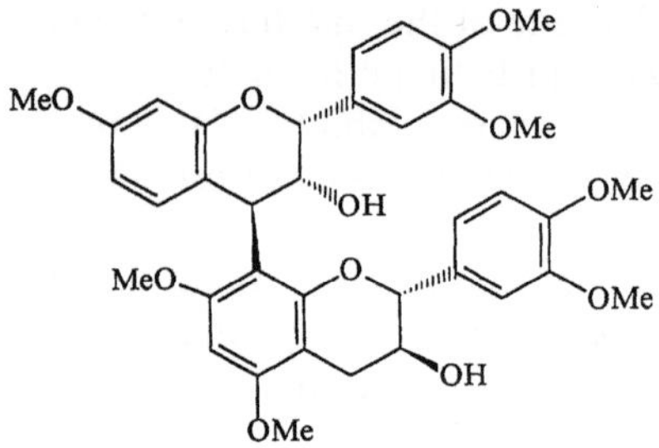

(85)

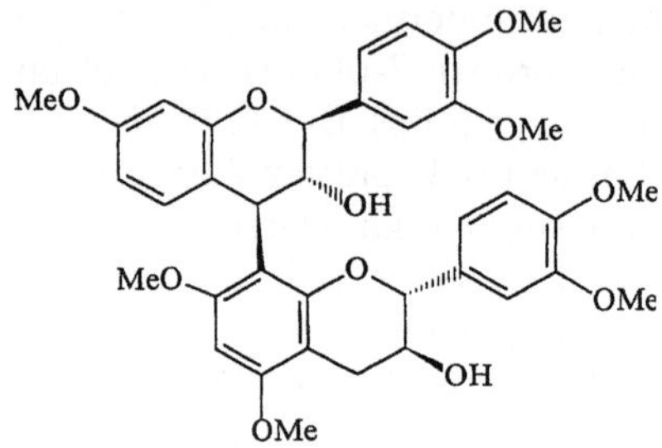

(86)

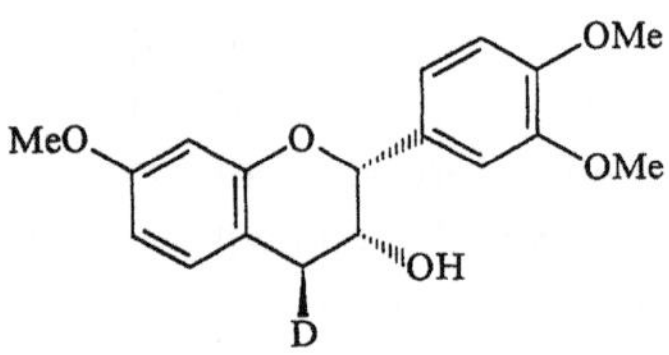

(87)

presumably indicates that delivery of hydride ion occurs from a complex between the reducing agent and the C-ring heterocyclic oxygen lone pair *trans* to the 2-aryl group, such transfer being most readily facilitated in an A-conformer (*36*) of type (**88**).

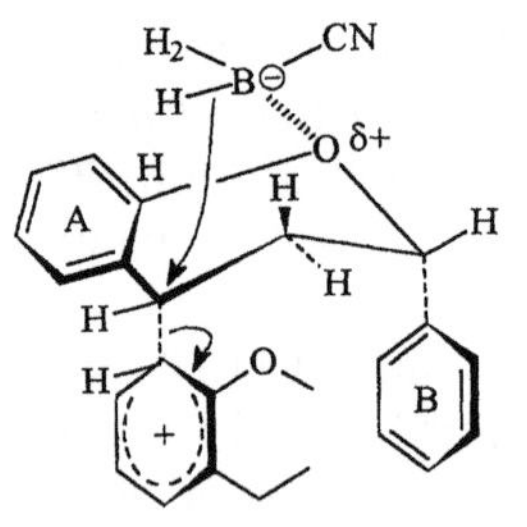

(88)

The potential of this development for the structural elucidation of the proanthocyanidin condensed tannins, especially the 5-deoxy analogues, from important commercial sources is clear. In addition, the method facilitates the ready definition of the absolute configuration of the chain-terminating flavan-3-ol moiety in 5-deoxyoligoflavanoids, especially in

References, pp. 59–67

view of the demonstration that these units may also comprise *ent*-catechin and *ent*-epicatechin (*88*), (*89*).

3.2. A-Type Proanthocyanidins

The double interflavanyl linkage in A-type proanthocyanidins introduces a high degree of conformational stability which culminates in high-quality and unequivocal NMR spectra conspicuously free of the effects of dynamic rotational isomerism at the dimeric level. Compounds of this class are readily recognizable from the characteristic AB-doublet ($^3J_{3,4} = 3-4$ Hz) of the C-ring protons in the heterocyclic region of their ^{1}H NMR spectra (*90*), and may possess either (2α,4α)- or (2β,4β)-double interflavanyl bonds. Two fundamental structural problems, *i.e.* establishment of the mode of linkage of the C- to the D-ring, and assignment of the absolute configuration at the stereocentres of the F-ring, have limited progress in this field. These and related problems have hitherto been approached by means of exotic spectroscopic methods (*74, 91–94*). This has prompted a search for a simpler and general chemical method that is based upon the reductive cleavage of the acetal functionality of A-type proanthocyanidins. The potential to address these problems by reduction of either of the C–O acetal bonds was demonstrated (*95*) for the known procyanidins A-1 (**89**) and A-2 (**90**), available from the skins of mature peanuts (*Arachus hypogea*) (*96*), by using $Na(CN)BH_3$ in TFA. The readily accessible hepta-*O*-methyl ethers (**91**) and (**92**) were selected as model compounds with a view to using the *O*-substituents of the D-ring as probes for anticipated much simplified ^{1}H NMR studies.

Separate treatment of the hepta-*O*-methylprocyanidins A-1 (**91**) and A-2 (**92**) with $Na(CN)BH_3$ in TFA for 1.5 h at 0°C (Scheme 7) provided mixtures comprising the starting materials and, as anticipated from cleavage 'a', the tetrahydropyrano[2,3-*f*]chromene derivatives (**93**) (5.2%) and (**94**) (7%). The B-type procyanidin biflavanoids (**95**) and (**96**) envisaged from 'b' type cleavage were not obtained but instead, the respective monomeric units, *i.e.* tetra-*O*-methyl-*ent*-catechin (**97**) (4%) and tri-*O*-methylcatechin (**98**) (3.4%) from the A-1 derivative (**91**), and tetra-*O*-methyl-*ent*-catechin (**97**) (3%) and tri-*O*-methylepicatechin (**99**) (1.3%) from the A-2 derivative (**92**) were isolated.

Both carbon-oxygen bonds of the acetal functionality in the procyanidin A-1 (**91**) and A-2 (**92**) derivatives are thus susceptible to reductive cleavage under acidic conditions. This process is presumably triggered by random protonation of the acetal oxygens and concomitant

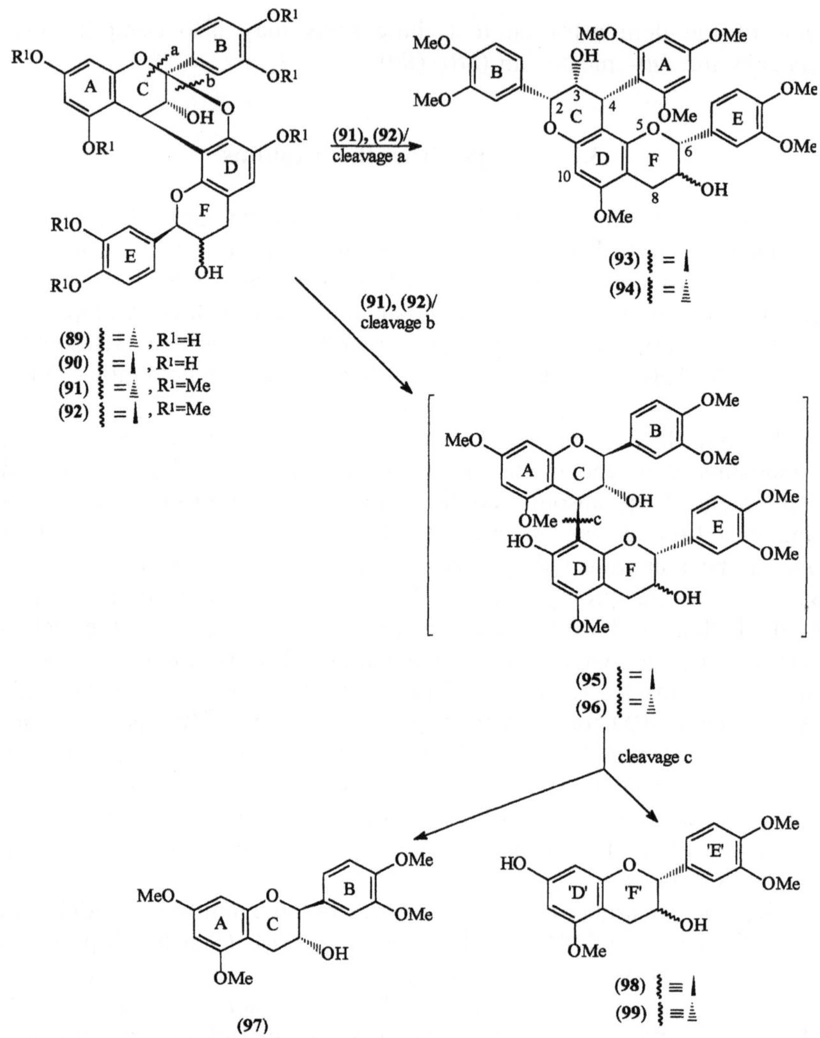

Scheme 7. Cleavage of the acetal functionality of procyanidin A-1 and A-2 permethylaryl ethers (**91**) and (**92**) with $Na(CN)BH_3$ in THF

delivery of the equivalent of hydride ion at the antibonding (σ^*) orbitals of the carbon-oxygen bonds in a predominant S_N2 manner. Such a transfer of hydride ion apparently occurs from a complex between the reducing agent and the *axial* C(3) (C-ring) oxygen lone pair, the proximity of the boron-hydrogen bonds to the backside of the acetal carbon being a prerequisite for reduction of either one of the acetal bonds. Reduction thus leads to 'inversion' of configuration at C(2)(C) of both B-type procyanidin intermediates (**95**) and (**96**), and of the tetrahydropyrano[2,3-*f*]chromene derivatives (**93**) and (**94**). The chemistry and the unequivocal structure elucidation, including assessment of absolute configuration at all the stereocentres of the latter class of compounds, are well understood (*28*), and facilitated confirmation of the absolute stereochemistry of ring F in the natural product derivatives (**91**) and (**92**).

Biflavanoids (**95**) and (**96**) are prone to facile cleavage of their interflavanyl bonds *via* protonation of the electron-rich phloroglucinol D-ring (*86, 87*) and attack of hydride ion at C(4)(C) (*83*) to give the *ent*-catechin derivative (**97**) from the ABC-unit and, respectively, the catechin (**98**) and epicatechin (**99**) derivatives from the DEF-moieties. The 'liberation' of the latter two chain terminating flavan-3-ol units unambiguously defines the D-ring oxygen that is involved in the acetal functionality of the parent compounds (**91**) and (**92**). It furthermore provides a powerful probe for addressing the hitherto unsolved problem of establishing the absolute configuration at the stereocentres of this moiety in naturally occurring A-type proanthocyanidins. The flavan-3-ol unit (**97**), albeit with inversed C(2) configuration, should facilitate the assignment of the absolute configuration at C(3) (C-ring) of the parent compounds (**91**) and (**92**), especially in view of the inability to differentiate between 3,4-*cis*- and 3,4-*trans*-configuration in these compounds on the basis of $^3J_{3,4}$ values (*74*). The mode of the C–C linkage between the constituent flavan-3-ol units in the A-type procyanidin, *e.g.* (4→6) or (4→8) is defined by the nature of the tetrahydropyranochromene (*28*), *i.e.* [2,3-*f*], [3,2-*g*] or [2,3-*h*], that is formed *via* reductive cleavage 'a'.

The protocol described here should thus contribute substantially towards a straightforward chemically orientated structural definition of the A-class proanthocyanidins.

4. Rearrangement of the Pyran Heterocycle of Oligomeric Proanthocyanidins

4.1. Introduction

Condensed tannins are often extracted and/or allowed to react at alkaline pH in the course of manufacture of speciality polymers such as tannin-based adhesives. These preparations invariably exhibit increased acidity and lower reactivity towards aldehydes than those obtained by neutral-solvent extraction (*97–99*), phenomena which have been attributed to the presence of catechinic acid-type rearrangement products (*97, 100*). With the exception of some reactions of monomeric flavan-3-ols (*97, 99–105*), studies of the base-catalyzed reactions of oligomeric proanthocyanidins during the pre-1988 era focussed mainly on flavan derivatives with "good" leaving groups at C(4) (*106–108*), the effects of external nucleophiles on intramolecular rearrangements and the lability of the interflavanyl bond and pyran ring at high pH values (*109, 110*),

4.2. Base-Catalyzed Pyran Ring Rearrangement of Oligomeric Proanthocyanidins

The natural occurrence and synthesis of a novel class of C-ring isomerized oligomeric flavanoids, termed phlobatannins, was demonstrated some twelve year ago (*111, 112*). These 3,4,9,10-tetrahydro-2*H*,8*H*-pyrano[2,3-*f*]chromenes, *e.g.* (**104**), are characterized by the 'liberated' resorcinol moieties from the A/C-ring arrangement of the parent biflavanoid, *e.g.* (**24**), and by the conspicuous absence of the effects of dynamic rotational isomerism in the ^{1}H NMR spectra of their permethylaryl ether diacetates at ambient temperatures.

Initial identification of the pyran-ring rearranged profisetinidins was followed by recognition of additional members of this class of oligoflavanoids from the heartwood of *Colophospermum mopane* (*42, 113–119*), *Guibourtia coleosperma* (*47, 114, 115, 120*), *Baikiaea plurijuga* (*47, 88, 115, 120*) *Julbernardia globiflora* (*88*), and the commercially important extract of the bark of *Acacia mearnsii* (*76*). Since the usual methods of differentiating regio-isomeric bi- and triflavanoids and the establishment of absolute configuration are less reliable for the phlobatannins, a concise synthetic protocol was developed to establish the complex structures of these natural products. The principles of such an approach are summarized in Scheme 8.

Biflavanoids, protected at 4-OH(E) in order to prevent the unwanted side-reactions that are associated with the formation of an E-ring quinone methide (*47, 120*), *e.g.* the fisetinidol-(4→8)-catechin profisetinidins (**100**) and (**101**), are susceptible to base-catalyzed cleavage of the C-ring with the formation of the B-ring quinone methides (**102**) and (**103**). Quinone methide (**102**) which is derived from the dimer with 3,4-*trans* (C-ring) configuration undergoes a highly stereoselective recyclization involving 7-OH(D) and the *re*-face at C(2) to give the tetrahydropyrano[2,3-*f*]-chromene (**104**). This process thus invariably leads from the 3,4-*trans* configuration in the parent biflavanoid (**100**) to

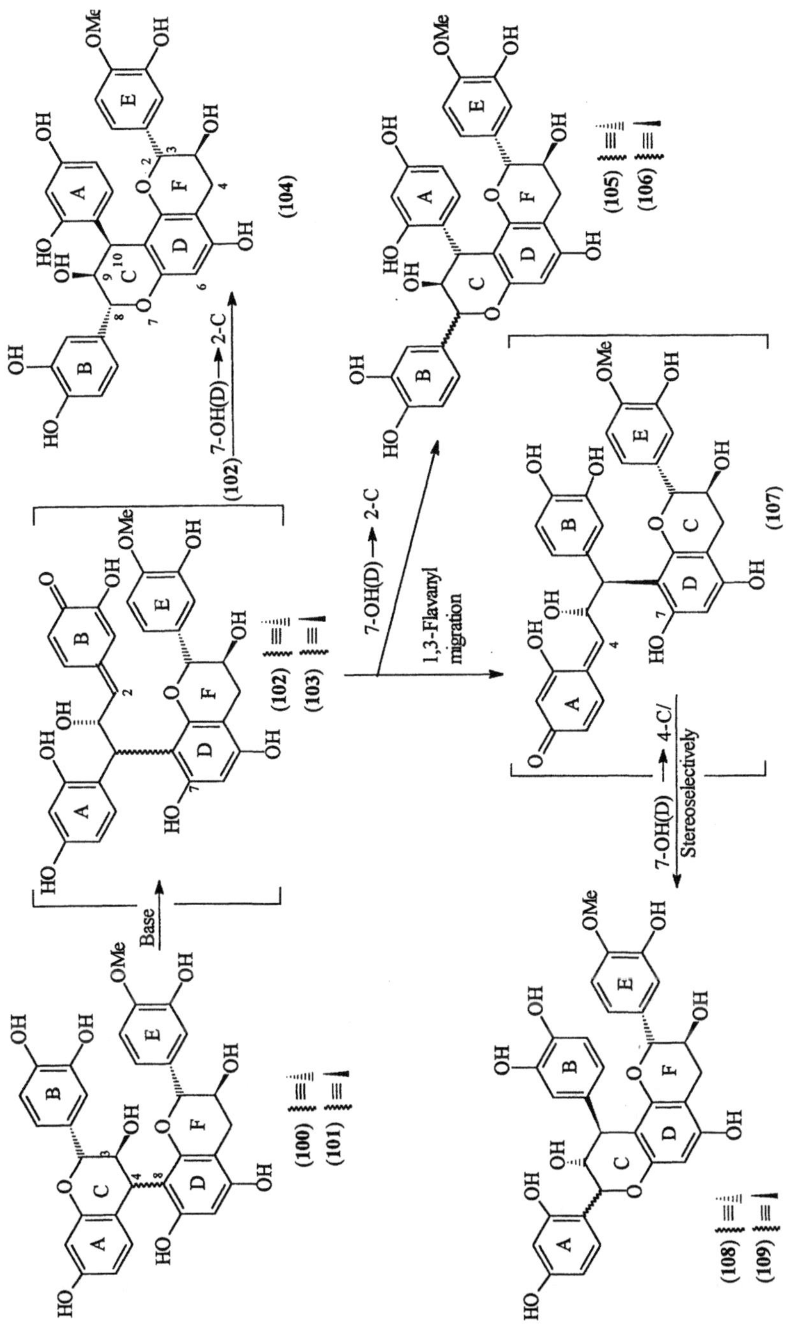

Scheme 8. Base-catalyzed pyran ring rearrangements of fisetinidol-(4α→8)- and (4β→8)-catechin 4-O (E-ring) methyl ethers (**100**) and (**101**)

the 9,10-*cis* arrangement in the phlobatannin (**104**). Besides its stereoselective recyclization involving 7-OH(D) and both the *re*- and *si*-faces at C(2) to give the tetrahydropyrano[2,3-*f*]chromenes (**105**) and (**106**), the quinone methide (**103**) is also susceptible to an unusual 1,3-migration of the catechin DEF-unit to give the A-ring quinone methide (**107**) (*47*, *120*). Stereoselective recyclization involving 7-OH(D) and C(4) then gives the tetrahydropyrano[2,3-*f*]chromenes (**108**) and (**109**) with interchanged resorcinol A- and pyrocatechol B-rings, and with inversed absolute configuration at C-9(C), compared to the arrangement

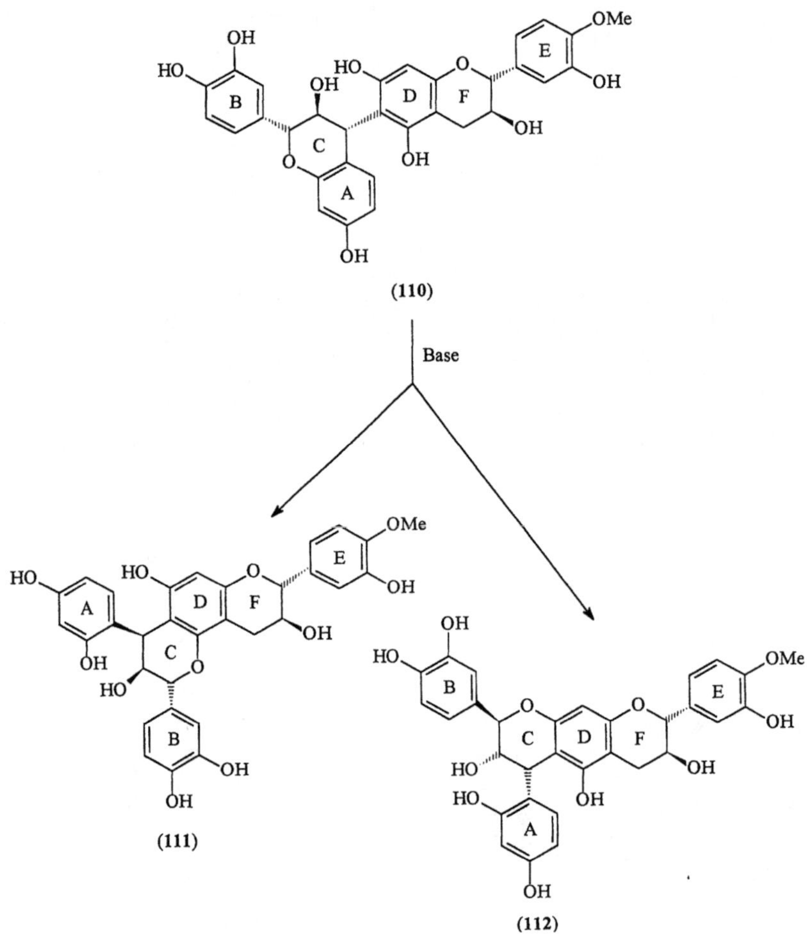

Scheme 9. Base-catalyzed pyran ring rearrangement of fisetinidol-(4α→6)-catechin 4-O (E-ring) methyl ether (**110**)

prevailing in the 'normal' analogues (**105**) and (**106**) (*47, 120, 121*). Quinone methides with phloroglucinol-type A-rings, *i.e.* those derived from procyanidin B-2 (**29**) (*75*) and B-3 (**32**) (*122*) additionally undergo 1,3-migration of this phloroglucinol moiety under the influence of the electron releasing D-ring, hence initiating the formation of a complex series of 2-flavanyl-4-aryl-3,4-dihydro-2*H*-benzopyrans (*75, 122*). Profisetinidins with (4→6)-interflavanyl linkages, *e.g.* (**110**) are transformed by base into the regio-isomeric tetrahydropyrano[2,3-*h*]- and [3,2-*g*]-chromenes (**111**) and (**112**) (Scheme 9) (*120*).

The aforementioned principles also govern the base-catalyzed C-ring isomerization of trimeric profisetinidins (*114, 116–119*), *e.g.* fisetinidol-(4α→6)-catechin-(8→4β)-fisetinidol (**113**). Analogues possessing constituent chain-extender units with 3,4-*cis*-stereochemistry [ABC unit in (**113**)] are similarly subject to extensive 1,3-migrations and thus to the formation of exceptionally complex reaction mixtures (*116–119*). This has led to the development of a more controlled synthesis that is based upon the repetitive formation of the interflavanyl bond and pyran ring rearrangement of the chain-extender unit under mild basic conditions (*123*). Thus, in contrast to the unrestrained course of the base-catalyzed C-ring rearrangement reactions of profisetinidin triflavanoids possessing 2,3-*trans*-3,4-*cis* flavanyl constituent units which result in exceptionally complex reaction mixtures, the stepwise construction of the dipyranochromene framework *via* sequential interflavanyl bond formation and pyran ring rearrangement permitted concise synthetic access to phlobatannins at the trimeric level.

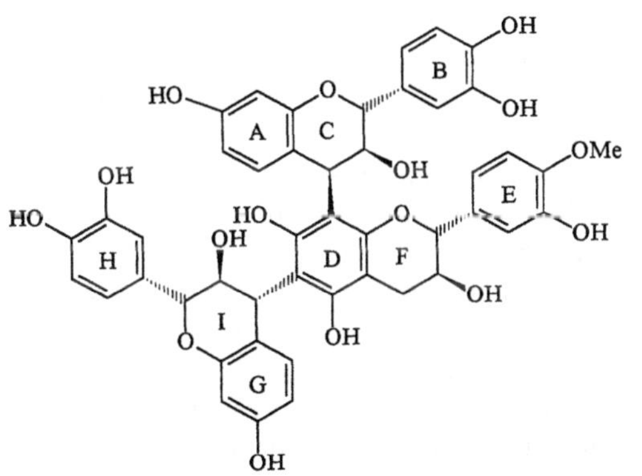

(113)

The susceptibility of the constituent flavanyl units of proanthocyanidins to intramolecular rearrangement *via* B-ring quinone methides under basic conditions was also demonstrated in an unusual dimerization-rearrangement reaction of catechin at pH 12 and 40°C (*124*).

Collectively, the work described in this section is of fundamental importance to an understanding of the chemistry of oligomeric proanthocyanidins in basic solution. It provides a basis for the commercial utilization of proanthocyanidins and also an understanding of the *post mortem* processes involved in the ageing of these biopolymers in wood and bark. The recognition of the phlobatannins also contributes to a rational explanation for the much reduced solubility of 'aged' proanthocyanidins in aqueous solvents. The phlobatannins all exhibit the characteristic structural features that are essential for the use of 'Momisa' extract in cold-setting adhesives and leather-tanning applications (*125*); thus their abundant presence in the bark extract (*76*) may well explain the industrial utility of this important renewable resource.

4.3. Acid-Catalyzed Rearrangement of Procyanidins and Prodelphinidins under Conditions of Thiolytic Cleavage

It was recently reported (*126*) that the application of thiolytic cleavage to study the condensed tannins from tree barks and nut shells did not provide quantitative yields of monomeric flavanyl cleavage products. Applications of thiolytic cleavage to the polymeric proanthocyanidins from pecan nut pith, known to be comprised of epigallocatechin, gallocatechin and epicatechin chain extender units in the approximate ratios of 5:2:1 with either catechin or gallocatechin as terminal units, consistently afforded significant amounts of phloroglucinol and a mixture of 1,3-dithiobenzyl-2,4,5,6-tetrahydroxyindane diastereomers (**120**).

Such a conversion is demonstrated in Scheme 10 for a typical prodelphinidin (**114**) with 2,3-*cis* configuration of the chain extender units. Thiolytic cleavage of (**114**) gives the 4β-benzylsulfanylepigallocatechin (**115**) which is protonated at the electron-rich phloroglucinol A-ring to afford intermediate (**116**) with a labile C(4)–C(10) bond which then ruptures under the influence of the electon-releasing benzylsulfanyl group. This is a unique process representing the equivalent of the cleaving of the interflavanyl bond under acidic conditions but under the influence of an external sulfur nucleophile. Rearrangement of the intermediate sulfonium ion (**117**) leads to the formation of the indane diastereomeric mixture (**118**) with a labile benzylic ether linkage which is cleaved, with the release of phloroglucinol, to carbocation (**119**). Reaction of the latter with, phenylmethane thiol affords the mixture of

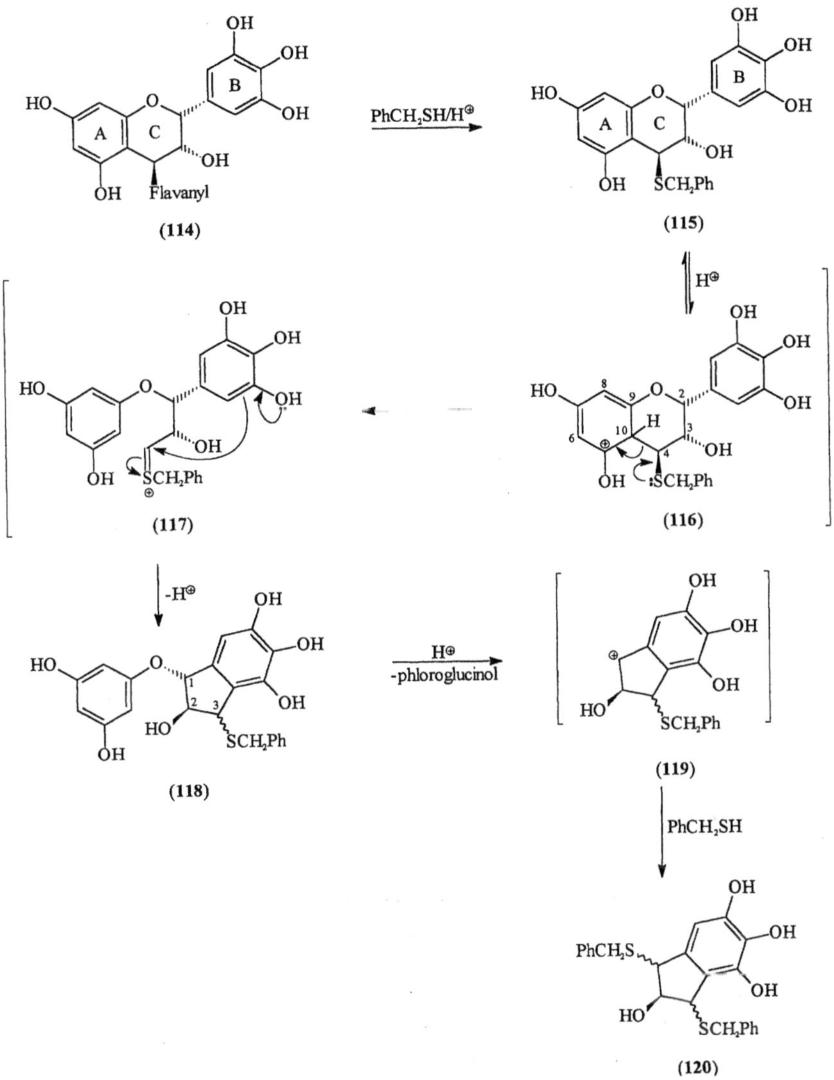

Scheme 10. Proposed route to the formation of phloroglucinol and indane diastereomers (**120**) during thiolysis of a prodelphinidin (**114**)

1,3-dibenzylsulfanyl-2,4,5,6-tetrahydroxyindane diastereomers (**120**). These results clearly invalidate the use of extended thiolysis to provide meaningful estimates of the molecular weight of polymeric proanthocyanidins. It also calls into question the use of thiolysis as a means of obtaining 'quantitative' information on the composition of mixed proanthocyanidin polymers.

These acid-catalyzed rearrangement reactions were also extended to a study in which pecan (*Caraya illinoensis*) nut pith tannins were reacted with phloroglucinol and acetic acid at 100°C for extended periods of time (*127*). Besides the formation of the anticipated 4-arylflavan-3-ols, these conditions also catalyze the formation of a unique series of novel [1]-benzofuro[2,3-*c*]chromenes, *e.g.* (**121**), and spirobidihydro[1]benzofurans, *e.g* (**122**). The formation of products of type (**121**) demonstrates the susceptibility of the C(3)-hydroxyl group in 4-arylflavan-3-ols to inversion of configuration under acidic conditions, while that of the spiro compounds of type (**122**) is a further manifestation of the lability of the C(4)–C(10) bond in condensed tannin constituent unit.

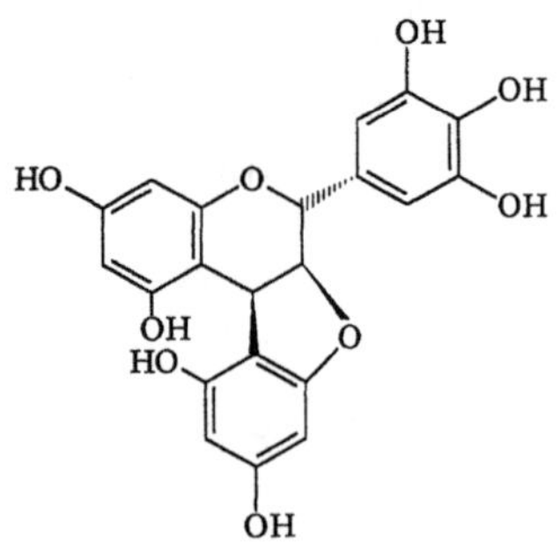

(121)

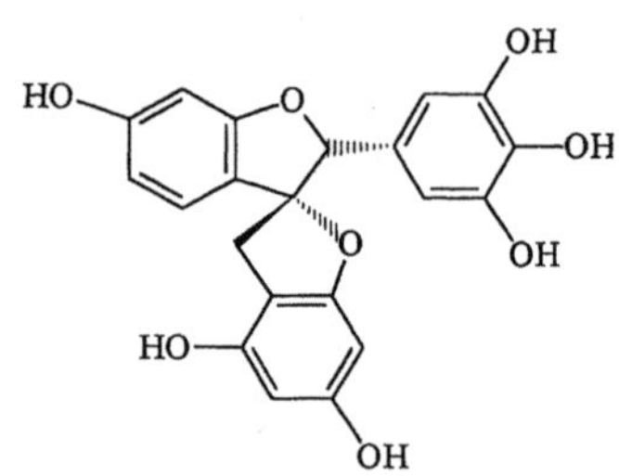

(122)

5. Conformational Analysis of Dimeric Proanthocyanidins

Conformational analysis of proanthocyanidin oligomers is in principle concerned with the conformation of the pyran heterocycles and with the phenomenon of conformational isomerism due to restricted rotation about the interflavanyl bond(s). Realization of the fact that the conformational itinerary of the heterocyclic rings involves a dynamic equilibrium between E- and A-conformers (*36*) had a huge impact in this field (*3*, *28*). It is generally being accepted that an understanding of the biological significance of the polymeric proanthocyanidins relies on insight into their complexation with other biopolymers. This had led to substantial efforts to improve comprehension of the interaction of polyflavanoids with proteins.

Much of this effort has been focussed on developing a more detailed understanding of the conformational preferences and flexibility of the proanthocyanidin polymers and their interaction with polypeptides. The conformational properties of the polyflavanoids have thus been studied by using a variety of molecular mechanics and molecular orbital computations (*7*, *10*, *14*) in combination with crystal structures (*11*, *12*), time-resolved fluorescence (*8*, *9*) as well as ^{1}H and ^{13}C NMR methods (*13*, *14*, *17*). Representative references to these techniques may be found in the papers listed in references (*14*)–(*20*), which in themselves are arguably the most authoritative reports recently published on this important branch of the chemistry of the proanthocyanidins. These results are summarized using the significant recent contributions of HATANO and HEMINGWAY (*19*).

NMR analysis of procyanidin B-1 (**27**) and B-3 (**32**) permitted full assignment of the proton and carbon resonances for both the more extended (**124**) and compact (**123**) conformers in the free phenolic form. In organic solvents the more extended rotamer (**124**) of procyanidin B-1 (**27**) is preferred over the more compact rotamer (10:7) but in water, the more compact rotamer dominates (10:2). When procyanidin B-3 (**32**) is dissolved in organic solvents, the more compact rotamer is slightly preferred (8:10). With water as solvent only trace proportions of the more extended rotamer are detected. In this solvent rotational conformation exchange is detected despite the observation of two distinct and sharp sets of signals for each rotamer. The heterocyclic ring of the ABC unit exists in an approximate half-chair conformation in each rotamer for both procyanidin B-1 (**27**) and B-3 (**32**). Coupling constants of the heterocyclic ring of the DEF moiety in both (**27**) and (**32**) indicate substantial axial orientation of the E-ring [see (**125**) and (**126**) for E- and A-conformers of the DEF unit of (**32**)]. Lineshape analysis of 3-H(F) indicated that the 'abnormal' coupling constants of the F-ring were indicative of a comparatively high-energy skewed-boat conformation for (**27**) and between a half-chair and a skewed-boat conformation for (**32**) rather than to E$\rightleftharpoons$A-conformational exchange which has hitherto been used to explain the smaller than anticipated coupling constants.

HATANO and HEMINGWAY (*19*) used NOE studies to assess the association of (+)-catechin (**5**) and procyanidin B-3 (**32**) with oligopeptides. These efforts focussing on the complexation of (poly)flavanoids with peptides containing proline residues in aqueous solutions revealed a site specific approach directed by hydrophobic interaction of the aromatic ring of (+)-catechin (**5**) and procyanidin B-3 (**32**) to conformationally accessible regions of peptides without strong preference for interaction with proline residues. The observed intermolecular NOE's indicating the

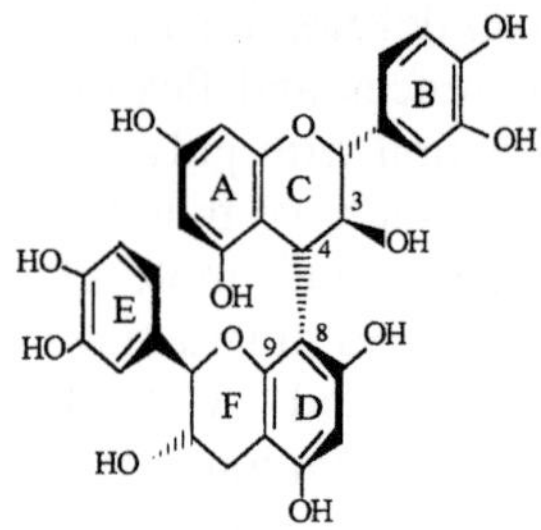

(123) [compact rotamer of (**32**)]
C_3-C_4-D_8-D_9 torsion angle , (-)

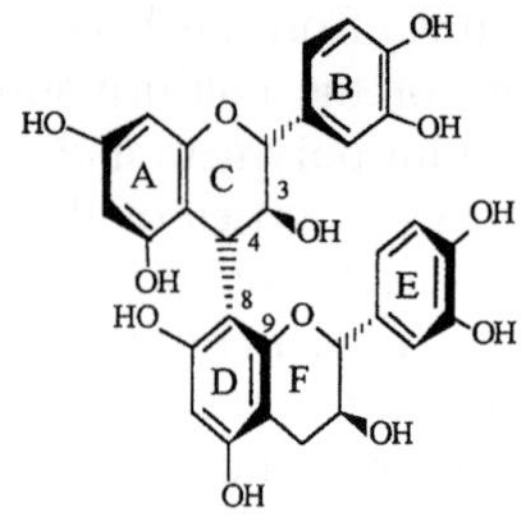

(124) [extended rotamer of (**32**)]
C_3-C_4-D_8-D_9 torsion angle , (+)

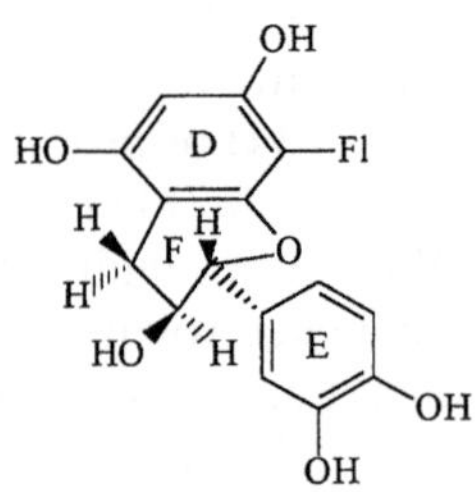

(125) : E-conformer

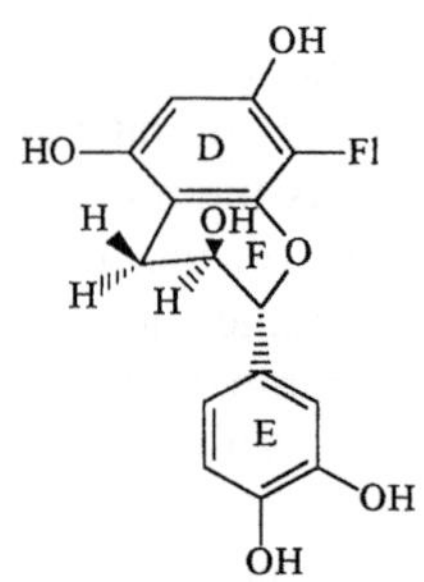

(126) : A-conformer

preferred sites in the association of (+)-catechin (**5**) and procyanidin B-3 (**32**) with the tetra-peptide, Gly-Pro-Gly-Gly are shown in (**127**) and (**128**), respectively.

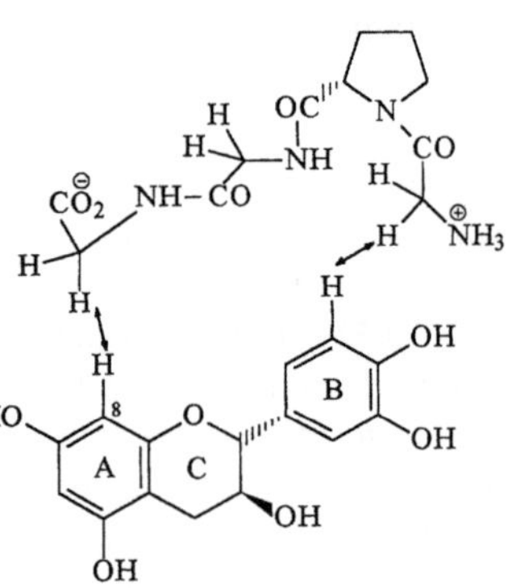

(127)

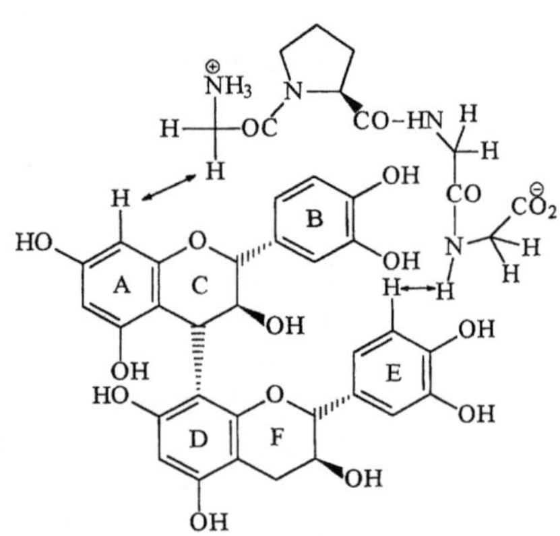

(128)

6. Conclusion

This review clearly demonstrates that considerable progress has been made to gain insight into the complex factors that govern the chemistry of the proanthocyanidin oligomers. It may be anticipated that the rapid advances that have been made in conformational analysis of these compounds will continue and will contribute towards understanding of the intricate principles governing the complexation of proanthocyanidins with other biomolecules.

Acknowledgements

We are grateful for the enthusiastic support of all our co-workers of the past 12 years. Financial support by the Foundation for Research Development, Pretoria and the "Sentrale Navorsingsfonds" of this University is gratefully acknowledged.

References

1. PORTER, L.J.: Flavans and Proanthocyanidins. In: The Flavonoids. Advances in Research since 1980. (J.B HARBORNE, ed.), p. 21. London, Chapman and Hall. 1988; PORTER, L.J.: Flavans and Proanthocyanidins. In: The Flavonoids. Advances in Research since 1986 (J.B. HARBORNE, ed.), p. 23. London, Chapman and Hall, 1994.
2. HEMINGWAY, R.W.: Biflavonoids and Proanthocyanidins. In: Natural Products of Woody Plants I (J.W. ROWE, ed.), p. 571. Berlin, Springer-Verlag, 1989.
3. FERREIRA, D., and R. BEKKER: Oligomeric Proanthocyanidins: Naturally-occurring *O*-heterocycles. Nat. Prod. Rep., **13**, 411 (1996).
4. ROUX, D.G., and D. FERREIRA: Structure and Function in the Biomimetic Synthesis of Linear, Angular and Branched Condensed Tannins. Pure & Appl. Chem., **54**, 2465 (1982).
5. HEMINGWAY, R.W., J.J. KARCHESY, and S.J. BRANHAM (eds.): Chemistry and Significance of Condensed Tannins, New York, Plenum Press, 1989.
6. HEMINGWAY, R.W., P.E. LAKS, and S.J. BRANHAM (eds.): Plant Polyphenols: Synthesis, Properties, Significance, New York, Plenum Press, 1992.
7. VISHWANADHAN, V.N., and W.L. MATTICE: Assessment by Molecular Mechanics of the Preferred Conformations of the Sixteen C(4)–C(6) and C(4)–C(8) Linked Dimers of (+)-Catechin and (−)-Epicatechin with Axial or Equatorial Dihydroxyphenyl Substituents at C(2). J. Chem. Soc., Perkin Trans. II, 739 (1987).
8. BERGMAN, W.R., M.D. BARKLEY, R.W. HEMINGWAY, and W.L. MATTICE: Heterogeneous Fluorescence Decay of (4→6)- and (4→8)-Linked Dimers of (+)-Catechin and (−)-Epicatechin as a Result of Rotational Isomerism. J. Am. Chem. Soc., **109**, 6614 (1987).
9. CHO, D., L.J. TIAN, L.J. PORTER, R.W. HEMINGWAY, and W.L. MATTICE: Variations in the Heterogeneity of the Decay of the Fluorescence in Six Procyanidin Dimers. J. Am. Chem. Soc., **112**, 4273 (1990).

10. STEYNBERG, J.P., E.V. BRANDT, and D. FERREIRA: Conformational Analysis of Oligomeric Flavanoids. Part 1. 4-Arylflavan-3-ols. J. Chem. Soc., Perkin Trans. II, 1569 (1991).

11. FRONCZEC, F.R., R.W. HEMINGWAY, G.W. MCGRAW, J.P. STEYNBERG, C.A. HELFER, and W.L. MATTICE: Crystal Structure, Conformational Analysis, and Molecular Dynamics of Tetra-*O*-methyl-(+)-catechin. Biopolymers, **33**, 275 (1993).

12. TOBIASON, F.L., F.R. FRONCZEC, J.P. STEYNBERG, E.C. STEYNBERG, and R.W. HEMINGWAY: Crystal Structure, Conformational Analyses, and Charge Density Distributions for *ent*-Epifisetinidol: An Explanation for Regiospecific Electrophilic Aromatic Substitution of 5-Deoxyflavans. Tetrahedron, **49**, 5927 (1993).

13. STEYNBERG, J.P., E.V. BRANDT, M.J.H. HOFFMANN, R.W. HEMINGWAY, and D. FERREIRA: Conformations of Proanthocyanidins. In: Plantpolyphenols: Synthesis, Properties, Significance (HEMINGWAY, R.W., P.E. LAKS, and S.J. BRANHAM ed.), p. 501, New York, Plenum Press, 1992.

14. STEYNBERG, J.P., E.V. BRANDT, D. FERREIRA, C.A. HELFER, W.L. MATTICE, D. GORNIK, and R.W. HEMINGWAY: Conformational Analysis of Oligomeric Flavanoids. Part 2. Methyl Ether Acetate Derivatives of Profisetinidins. Magn. Reson. Chem., **33**, 611 (1995).

15. BALAS, L., J. VERCAUTEREN, and M. LAQUERRE: 2D NMR Structure Elucidation of Proanthocyanidins: The Special Case of the Catechin-(4α-8)-catechin-(4α-8)-catechin Trimer. Magn. Reson. Chem., **33**, 85 (1995).

16. VIVAS, N., Y. GLORIES, I. PIANET, B. BARBE, and M. LAQUERRE: A Complete Structural and Conformational Investigation of Procyanidin A2 Dimer. Tetrahedron Lett., **37**, 2015 (1996).

17. HEMINGWAY, R.W., F.L. TOBIASON, G.W. MCGRAW, and J.P. STEYNBERG: Conformation and Complexation of Tannins: NMR Spectra and Molecular Search Modeling of Flavan-3-ols. Magn. Reson. Chem., **34**, 424 (1996).

18. De BRUYNE, T., L.A.C. PIETERS, R.A. DOMMISSE, H. KOLODZIEJ, V. WRAY, T. DOMKE, and A.J. VLIETINCK: Unambiguous Assignments of Free Dimeric Proanthocyanidin Phenols from 2D NMR. Phytochemistry, **43**, 265 (1996).

19. HATANO, T., and R.W. HEMINGWAY: Association of (+)-Catechin and Catechin-(4α→8)-catechin with Oligopeptides. Chem. Commun., 2537 (1996); Conformational Isomerism of Phenolic Procyanidins: Preferred Conformations in Organic Solvents and Water. J. Chem. Soc., Perkin Trans. II, 1035 (1997).

20. KHAN, M.L., E. HASLAM, and M.P. WILLIAMSON: Structure and Conformation of the Procyanidin B-2 Dimer. Magn. Reson. Chem., **35**, 854 (1997).

21. SALAH, N., N.J. MILLER, G. PARANGA, L. TIJBURG, G.P. BOLWELL, and C. RICEEVANS: Polyphenolic Flavanols as Scavengers of Aqueous-phase Radicals and as Chain-breaking Antioxidants. Arch. Biochem. Biophys, **322**, 339 (1995).

22. KANNER, J., E.W. FRANKEL, R. GRANIT, B. GERMAN, and J.E. KINSELLA: Natural Antioxidants in Grapes and Wine. J. Agric. Food Chem., **42**, 64 (1994).

23. FUHRMAN, B., A. LAVY, and M. AVIRAM: Consumption of Red Wine with Meals Reduces the Susceptibility of Human Plasma and Low-density Lipoprotein to Lipid-peroxidation. Am. J. Clin. Nutr., **61**, 549 (1995).

24. FRANKEL, E.N., A.L. WATERHOUSE, and P.L. TEISSEDRE: Principal Phenolic Phytochemicals in Selected Californian Wines and their Antioxidant Activity in Inhibiting Oxidation of Low-density Lipoproteins. J. Agric. Food Chem., **43**, 890 (1995).

25. FRANKEL, E.N., J. KANNER, J.B. GERMAN, E. PARKS, and J.E. KINSELLA: Inhibition of Oxidation of Human Low-density Lipoprotein by Phenolic Substances in Red Wine. Lancet, **341**, 454 (1993).

26. RUF, J.C., J.L. BERGER, and S. RENAUD: Platelet Rebound Effect of Alcohol-withdrawal and Wine Drinking in Rats—Relation to Tannins and Lipid-peroxidation. Arterioscler. Thromb. Vasc. Biol., **15**, 140 (1995).
27. ROUX, D.G., and D. FERREIRA: The Direct Biomimetic Synthesis, Structure and Absolute Configuration of Angular and Linear Condensed Tannins. Fortschr. Chem. organ. Naturstoffe, **41**, 47 (1982).
28. FERREIRA, D., J.P. STEYNBERG, D.G. ROUX, and E.V. BRANDT: Diversity of Structure and Function in Oligomeric Flavanoids. Tetrahedron, **48**, 1743 (1992).
29. FERREIRA, D., J.P. STEYNBERG, J.F.W. BURGER, and B.C.B. BEZUIDENHOUDT: Synthesis and Base-catalyzed Transformations of Proanthocyanidins. In: Recent Advances in Phytochemistry. (STAFFORD, H.A., and R.K. IBRAHIM eds.), p. 255. New York, Plenum Press, 1992.
30. DREWES, S.E., D.G. ROUX, S.H. EGGERS, and J. FEENEY: Three Diastereomeric 4,6-Linked Bileucofisetinidins from the Heartwood of *Acacia mearnsii*. J. Chem. Soc. (C), 1217 (1967).
31. DREWES, S.E., D.G. ROUX, H.M. SAAYMAN, S.H. EGGERS, and J. FEENEY: Some Stereochemically Identical Biflavanols from the Bark Tannins of *Acacia mearnsii*. J. Chem. Soc. (C), 1302 (1967).
32. DU PREEZ, I.C., A.C. ROWAN, D.G. ROUX, and J. FEENEY: Hindered Rotation about the sp^2–sp^3 Hybridized C–C Bond Between Flavanoid Units in Condensed Tannins. Chem. Commun., 315 (1971).
33. FLETCHER, A.C., L.J. PORTER, and E. HASLAM: Hindered Rotation and Helical Structures in Natural Procyanidins. J. Chem. Soc., Chem. Commun., 627 (1976).
34. FLETCHER, A.C., L.J. PORTER, E. HASLAM, and R.J. GUPTA: Plant Proanthocyanidins. Part 3. Conformational and Configurational Studies of Natural Procyanidins. J. Chem. Soc., Perkin Trans. 1, 1628 (1977).
35. BROWN, B.R., and M.R. SHAW: Reactions of Flavonoids and Condensed Tannins with Sulphur Nucleophiles. J. Chem. Soc., Perkin Trans. 1, 2036 (1974).
36. PORTER, L.J., R.Y. WONG, M. BENSON, B.G. CHAN, V.N. VISHWANADHAN, R.D. GANDOUR, and W.L. MATTICE: Conformational Analysis of Flavans. ^{1}H NMR and Molecular Mechanical (MM2) Studies of the Benzopyran Ring of 3′,4′,5,7-Tetrahydroxyflavan-3-ols: The Crystal and Molecular Structure of the Procyanidin: (2*R*,3*S*,4*R*)-3′,4′,5,7-Tetramethoxy-4-(2,4,6-trimethoxyphenyl)-flavan-3-ol. J. Chem. Res. (M), 830; (S), 86 (1986).
37. MALAN, J.C.S., J.A. STEENKAMP, D.A. YOUNG, and D. FERREIRA: Oligomeric Flavanoids. Part 13. Synthesis of Profisetinidins Based on (−)-Robinetinidol and (+)-Epifisetinidol. Tetrahedron, **45**, 7859 (1989).
38. COETZEE, J., J.P. STEYNBERG, P.J. STEYNBERG, E.V. BRANDT, and D. FERREIRA: Oligomeric Flavanoids. Part 18. Dimeric Prorobinetinidins from *Robinia pseudacacia*. Tetrahedron, **51**, 2339 (1995).
39. HASLAM, E.: Biogenetically Patterned Synthesis of Procyanidins. J. Chem. Soc., Chem. Commun., 594 (1974).
40. BOTHA, J.J., D. FERREIRA, and D.G. ROUX: Condensed Tannins: Direct Synthesis, Structure, and Absolute Configuration of Four Biflavanoids from Black Wattle Bark ('Mimosa') Extract. J. Chem. Soc., Chem. Commun., 700 (1978); Synthesis of Condensed Tannins. Part 4. A Direct Biomimetic Approach to [4,6]- and [4,8]-Biflavanoids. J. Chem. Soc., Perkin Trans. 1, 1235 (1981).
41. STEYNBERG, P.J., J.P. STEYNBERG, E.V. BRANDT, D. FERREIRA, and R.W. HEMINGWAY: Oligomeric Flavanoids. Part 26. Structure and Synthesis of the First Profisetinidins with Epifisetinidol Constituent Units. J. Chem. Soc., Perkin Trans. 1, 1943 (1997).

42. MALAN, J.C.S., D.A. YOUNG, J.P. STEYNBERG, and D. FERREIRA: Oligomeric Flavanoids. Part 10. Structure and Synthesis of the First Tetrahydropyrano[2,3-*g*]chromenes Related to (4,6)-Bis-(−)-fisetinidol Profisetinidins. J. Chem. Soc., Perkin Trans. 1, 227 (1990).
43. FOO, L.Y.: A Novel Pyrogallol A-ring Proanthocyanidin Dimer from *Acacia melanoxylon*. J. Chem. Soc., Chem. Commun., 236 (1986).
44. MALAN, E., and A. SIREEPARSAD: The Structure and Synthesis of the First Dimeric Proteracacinidins from *Acacia galpinii*. Phytochemistry, **38**, 237 (1995).
45. CLARK-LEWIS, J.W.: Flavan Derivatives. XXI. Nuclear Magnetic Resonance Spectra, Configuration, and Conformation of Flavan Derivatives. Aust. J. Chem., **21**, 2059 (1968).
46. VAN DER WESTHUIZEN, J.H., D. FERREIRA, and D.G. ROUX: Synthesis of Condensed Tannins. Part 2. Synthesis by Photolytic Rearrangement, Stereochemistry and Circular Dichroism of the First 2,3-*cis*-3,4-*cis*-4-Arylflavan-3-ols. J. Chem. Soc., Perkin Trans. 1, 1220 (1981).
47. STEYNBERG, J.P., J.F.W. BURGER, D.A. YOUNG, E.V. BRANDT, J.A. STEENKAMP, and D. FERREIRA: Oligomeric Flavanoids. Part 4. Base-catalyzed Conversions of (−)-Fisetinidol-(+)-catechin Profisetinidins with 2,3-*trans*-3,4-*cis*-Flavan-3-ol Constituent Units. J. Chem. Soc., Perkin Trans. 1, 3331 (1988).
48. BOTHA, J.J., D.A. YOUNG, D. FERREIRA, and D.G. ROUX: Synthesis of Condensed Tannins, Part 1. Stereoselective and Stereospecific Synthesis of Optically Pure 4-Arylflavan-3-ols, and Assessment of Their Absolute Stereochemistry at C-4 by Means of Circular Dichroism. J. Chem. Soc., Perkin Trans. 1, 1213 (1981).
49. BARRETT, M.W., W. KLYNE, P.M. SCOPES, A.C. FLETCHER, L.J. PORTER, and E. HASLAM: Plant Proanthocyanidins. Part 6. Chiroptical Studies. Part 95. Circular Dichroism of Procyanidins. J. Chem. Soc., Perkin Trans. 1, 2375 (1979).
50. De ANGELIS, G.G., and W.C. WILDMAN: Circular Dichroism Studies–1. A Quadrant Rule for the Optically Active Aromatic Chromophore in Rigid Polycyclic Systems. Tetrahedron, **25**, 5099 (1968).
51. YOUNG, D.A., A. CRONJÉ, A.L. BOTES, D. FERREIRA, and D.G. ROUX: Synthesis of Condensed Tannins. Part 14. Biflavanoids as Synthons. The Acid-induced 'Phlobaphene' Reaction. J. Chem. Soc., Perkin Trans. 1, 2521 (1985).
52. STEYNBERG, P.J., R.J.J. NEL, H. VAN RENSBURG, B.C.B. BEZUIDENHOUDT, and D. FERREIRA: Oligomeric Flavanoids. Part 27. Interflavanyl Bond Formation in Procyanidins under Neutral Conditions. Tetrahedron, **54**, 8153 (1998).
53. TROST, B.M., and E. MURAYAMA: Dimethyl(methylthio)sulfonium Fluoroborate. A Chemoselective Initiator for Thionium Induced Cyclizations. J. Am. Chem. Soc., **103**, 6529 (1981).
54. TROST, B.M., and T. SATO: Dimethyl(methylthio)sulfonium Tetrafluoroborate Initiated Organometallic Additions to and Macrocyclizations of Thioketals. J. Am. Chem. Soc., **107**, 719 (1985).
55. BARRETT, A.G.M., B.C.B. BEZUIDENHOUDT, A.R. HOWELL, A.C. LEE, and M.A. RUSSEL: Redox Glycosylation *via* Thionoester Intermediates. J. Org. Chem., **54**, 2275 (1989).
56. DELCOUR, J.A., D. FERREIRA, and D.G. ROUX: Synthesis of Condensed Tannins. Part 9. The Condensation Sequence of Leucocyanidin with (+)-Catechin and with the Resultant Procyanidins. J. Chem. Soc., Perkin Trans. 1, 1711 (1983).
57. DREWES, S.E., and A.H. ILSLEY: Dioxan-linked Biflavanoid from the Heartwood of *Acacia mearnsii*. J. Chem. Soc. (C), 1302 (1969).
58. YOUNG, D.A., D. FERREIRA, and D.G. ROUX: Synthesis of Condensed Tannins. Part 10. 'Dioxan-linked' Profisetinidins. J. Chem. Soc., Perkin Trans. 1, 2031 (1983).

59. Foo, L.Y.: Isolation of [4-O-4]linked Biflavanoids form *Acacia melanoxylon*. First Examples of a New Class of Single Ether Linked Proanthocyanidin Dimers. J. Chem. Soc., Chem. Commun., 1505 (1989).
60. Malan, E., A. Sireeparsad, J.F.W. Burger, and D. Ferreira: A Novel Doubly-linked Proteracacinidin Analogue from *Acacia caffra*. Tetrahedron Lett., **35**, 7415 (1994).
61. Coetzee, J., E. Malan, and D. Ferreira: Oligomeric Flavanoids. Part 28. Structure and Synthesis of Ether-linked [4-O-3]Bis-teracacinidins, a Novel Class of Naturally Occurring Proanthocyanidins. J. Chem. Res. (M), 2287; (S), 526 (1998).
62. Coetzee, J., E. Malan, and D. Ferreira: Oligomeric Flavanoids. Part 29. Structure and Synthesis of Novel Ether-linked [4-O-4]Bis-teracacinidins. Tetrahedron, **54**, 9153 (1998).
63. Young, D.A., D. Ferreira, D.G. Roux, and W.E. Hull: Synthesis of Condensed Tannins. Part 15. Structure of Natural 'Angular' Profisetinidin Tetraflavanoids. Asymmetric Induction During Synthesis. J. Chem. Soc., Perkin Trans. 1, 2529 (1985).
64. Young, D.A., H. Kolodziej, D. Ferreira, and D.G. Roux: Synthesis of Condensed Tannins. Part 16. Stereochemical Differentiation of the First 'Angular' (2*S*,3*R*)-Profisetinidin Tetraflavanoids from *Rhus lancea* (Karee) and the Varying Dynamic Behaviour of Their Derivatives. J. Chem. Soc., Perkin Trans. 1, 2537 (1985).
65. Young, D.A., D. Ferreira, and D.G. Roux: Stereochemistry and Dynamic Behaviour of Some Synthetic 'Angular' Tetraflavanoid Derivatives. J. Polym. Sci., Part A: Polym. Chem., **24**, 835 (1986).
66. Brandt, E.V., D.A. Young, H. Kolodziej, D. Ferreira, and D.G. Roux: Cycloconformations of Two Tetraflavanoid Condensed Tannins. J. Chem. Soc., Chem. Commun., 913 (1986).
67. Brandt, E.V., D.A. Young, D. Ferreira, and D.G. Roux: Synthesis of Condensed Tannins. Part 20. Cycloconformations and Conformational Stability Among Derivatives of 'Angular' Tetraflavanoid Profisetinidins. J. Chem. Soc., Perkin Trans. 1, 2353 (1987).
68. Malan, E., A. Sireeparsad, E. Swinny, and D. Ferreira: The Structure and Synthesis of a 7,8,4′-Trihydroxyflavan-epioritin dimer from *Acacia caffra*. Phytochemistry, **44**, 529 (1997).
69. Hatano, T., A. Yamashita, T. Hashimoto, H. Ito, N. Kubo, M. Yoshiyama, S. Shimura, Y. Itoh, T. Okuda, and T. Yoshida: Flavan Dimers with Lipase Inhibitory Activity from *Cassia nomane*. Phytochemistry, **46**, 893 (1997).
70. Malan, E., E. Swinny, D. Ferreira, and P.J. Steynberg: The Structure and Synthesis of Proguibourtinidins from *Cassia abbreviata*. Phytochemistry, **41**, 1209 (1996).
71. Pierre, M-C., C. Chéze, and J. Vercauteren: Deuterium Labeled Procyanidin Synthesis. Tetrahedron Lett., **38**, 5639 (1997).
72. Yoneda, S., H. Kawamota, and F. Nakatsubo: Synthesis of High Molecular Mass Condensed Tannin by Cationic Polymerization of Flavan-3,4-carbonate. J. Chem. Soc., Perkin Trans. 1, 1025 (1997).
73. Morimoto, S., G. Nonaka, and I. Nishioka: Tannins and Related Compounds. LIX. Aesculitannins, Novel Proanthocyanidins with Doubly-bonded Structures from *Aesculus hippocastanum* L. Chem. Pharm. Bull., **35**, 4717 (1987).
74. Cronjé, A., J.F.W. Burger, E.V. Brandt, H. Kolodziej, and D. Ferreira: Assessment of 3,4-*trans*- and 3,4-*Cis* Relative Configuration in the A-series of (4,8)-Linked Proanthocyanidins. Tetrahedron Lett. **31**, 3789 (1990).
75. Burger, J.F.W., H. Kolodziej, R.W. Hemingway, J.P. Steynberg, D.A. Young, and D. Ferreira: Oligomeric Flavanoids. Part 15. Base-catalyzed Pyran Rearrangements of

Procyanidin B-2, and Evidence for the Oxidative Transformation of B- to A-type Proanthocyanidins. Tetrahedron, **46**, 5733 (1990).

76. CRONJÉ, A., J.P. STEYNBERG, E.V. BRANDT, D.A. YOUNG, and D. FERREIRA: Oligomeric Flavanoids. Part 16. Novel Prorobinetinidins and the First A-type Proanthocyanidin with 5-Deoxy A- and a 3,4-*Cis* C-ring from the Maiden Investigation of Commercial Wattle Bark Extract. J. Chem. Soc., Perkin Trans. 1, 2467 (1993).

77. KOLODZIEJ, H., D. FERREIRA, G. LEMIERÉ, T. De BRUYNE, L. PIETERS, and A.J. VLIETINCK: On the Nomenclature of Oligoflavanoids with an A-type Unit. J. Nat. Prod., **56**, 1199 (1993).

78. MUSSO, H.: In Oxidative Coupling of Phenols (TAYLOR, W.I., BATTERSBY, A.R., eds.), p. 75. New York, Marcel Dekker, Inc. 1967.

79. BETTS, M.J., B.R. BROWN, P.E. BROWN, and W.T. PIKE: Degradation of Condensed Tannins: Structure of the Tannin from Common Heather. Chem. Commun., 1110 (1967).

80. THOMPSON, R.S., D. JACQUES, E. HASLAM, and R.J. TANNER: Plant Proanthocyanidins. Part 1. Introduction; the Isolation, Structure, and Distribution in Nature of Plant Procyanidins. J. Chem. Soc., Perkin Trans. 1, 1387 (1972).

81. FOO, L.Y., and L.J. PORTER: Prodelphinidin Polymers: Definition of Structural Units. J. Chem. Soc., Perkin Trans. 1, 1186 (1978).

82. STEYNBERG, P.J., J.P. STEYNBERG, B.C.B. BEZUIDENHOUDT, and D. FERREIRA: Cleavage of the Interflavanyl Bond in 5-Deoxy (A-ring) proanthocyanidins. J. Chem. Soc., Chem. Commun., 31 (1995).

83. STEYNBERG, P.J., J.P. STEYNBERG, B.C.B. BEZUIDENHOUDT, and D. FERREIRA: Oligomeric Flavanoids. Part 19. Reductive Cleavage of the Interflavanyl Bond in Proanthocyanidins. J. Chem. Soc., Perkin Trans. 1, 3005 (1995).

84. LANE, C.F.: Sodium Cyanoborohydride, a Highly Selective Reducing Agent for Organic Functional Groups. Synthesis, (3), 135 (1975).

85. MCGRAW, G.W., and R.W. HEMINGWAY: Electrophilic Aromatic Substitution of Catechin. Bromination and Benzylation. J. Chem. Soc., Perkin Trans. 1, 973 (1982).

86. BROWN, A.G., W.B. EYTON, A. HOLMES, and W.D. OLLIS: Identification of the Thearubigenins as Polymeric Proanthocyanidins. Nature, **221**, 742 (1969).

87. BAERT, J.E., T.H. LILLEY, and E. HASLAM: Polyphenol Interactions. Part 2. Covalent Binding of Procyanidins to Proteins During Acid-catalyzed Decomposition: Observations on Some Polymeric Proanthocyanidins. J. Chem. Soc., Perkin Trans. 2, 1439 (1985).

88. STEYNBERG, P.J., J.F.W. BURGER, B.C.B. BEZUIDENHOUDT, J.P. STEYNBERG, M.S. VAN DYK, and D. FERREIRA: The First Natural Condensed Tannins with (−)-Catechin 'Terminal' Units. Tetrahedron Lett., **31**, 2059 (1990).

89. DELLE MONACHE, F., F. FERRARI, A. POCE-TUCCI, and G.B. MARINI-BETTOLLO: Catechins with (+)-Epi-configuration in Nature. Phytochemistry, **11**, 2333 (1972).

90. JACQUES, D., E. HASLAM, G.R. BEDFORD, and G. GREATBANKS: Plant Proanthocyanidins. Part II. Proanthocyanidin−A2 and its Derivatives. J. Chem. Soc., Perkin Trans. 1, 2663 (1974).

91. KOLODZIEJ, H., M.J. SAKAR, J.F.W. BURGER, R. ENGELSHOWE, and D. FERREIRA: A-Type Proanthocyanidins from *Prunus spinosa*. Phytochemistry, **30**, 2042 (1991).

92. GONZALEZ, A.G., A.C. IRIZAR, A.G. RAVELO, and M.F. FERNANDEZ: Type-A Proanthocyanidins from *Prunus spinosa*. Phytochemistry, **31**, 1432 (1992).

93. BALDE, A.M., T. De BRUYN, L. PIETERS, H. KOLODZIEJ, D. VAN DEN BERGHE, M. CLAEYS, and A. VLIETINCK: Oligomeric Proanthocyanidins Possessing a Doubly Linked Structure from *Pavetta owariensis*. Phytochemistry, **38**, 719 (1995).

94. Balde, A.M., T. De Bruyne, L. Pieters, H. Kolodziej, D. van den Berghe, M. Claeys, and A. Vlietinck: Tetrameric Proanthocyanidins Containing a Double Interflavanoid (A-Type) Linkage from *Pavetta owariensis*. Phytochemistry, **40**, 933 (1995).
95. Steynberg, P.J., A. Cronjé, J.P. Steynberg, B.C.B. Bezuidenhoudt, E.V. Brandt, and D. Ferreira: Oligomeric Flavanoids. Part 25. Cleavage of the Acetal Functionality in A-Type Proanthocyanidins. Tetrahedron, **53**, 2591 (1997).
96. Karchesy, J.J., and R.W. Hemingway: Condensed Tannins: (4β→8; 2β→O→7)-Linked Procyanidins in *Arachis hypogea* L. J. Agric. Food Chem., **34**, 966 (1986).
97. Sears, K.D., R.L. Casebier, H.L. Hergert, G.H. Stout, and L.E. McCandlish: The Structure of Catechinic Acid. A Base Rearrangement Product of Catechin. J. Org Chem., **39**, 3244 (1974).
98. Herrick, F.H.: Chemistry and Utilization of Western Hemlock Bark Extractives. J. Agric. Food Chem. **28**, 228 (1980).
99. Kiatgrajai, P., J.D. Wellons, L. Gollob, and J.D. White: Kinetics of Epimerization of (+)-Catechin and its Rearrangement to Catechinic Acid. J. Org. Chem., **47**, 2910 (1982).
100. Courbat, P., A. Weith, A. Albert, and A. Pelter: Contribution to the Study of the Behaviour of Catechin in Alkaline Medium. Helv. Chim. Acta, **60**, 1665 (1977).
101. Foo, L.Y., and L.J. Porter: Synthesis and Conformation of Procyanidin Diastereomers. J. Chem. Soc., Perkin Trans. 1, 1535 (1985).
102. Freudenberg, K., and L. Purrmann: Raumisomere Catechine, III. Chem. Ber., **56**, 1185 (1923).
103. Freudenberg, K., and L. Purrmann: Raumisomere Catechine, IV. Liebigs Ann. Chem., **437**, 274 (1924).
104. Kennedy, J.A., M.H.G. Munro, H.K.J. Powell, L.J. Porter, and L.Y. Foo: The Protonation Reactions of Catechin, Epicatechin and Related Compounds. Aust. J. Chem., **37**, 885 (1984).
105. Mehta, P.P., and W.B. Whalley: The Stereochemistry of Some Catechin Derivatives. J. Chem. Soc., 5327 (1963).
106. Hemingway, R.W., and L.Y. Foo: Condensed Tannins: Quinone-methide Intermediates in Procyanidin Synthesis. J. Chem. Soc., Chem. Commun., 1035 (1983).
107. Foo, L.Y., and R.W. Hemingway: Condensed Tannins: Synthesis of the First 'Branched' Procyanidin Trimer. J. Chem. Soc., Chem. Commun., 85 (1984).
108. Hemingway, R.W., and P.E. Laks: Condensed Tannins: A Proposed Route to 2*R*,3*R*-(2,3-*cis*)-Proanthocyanidins, J. Chem. Soc., Chem. Commun., 746 (1985).
109. Laks, P.E., and R.W. Hemingway: Condensed Tannins: Base-catalyzed Reactions of Polymeric Procyanidins with Toluene-α-thiol. Lability of the Interflavonoid Bond and Pyran Ring. J. Chem. Soc., Perkin Trans. 1, 465 (1987).
110. Laks, P.E., R.W. Hemingway, and A.H. Conner: Condensed Tannins. Base-catalyzed Reactions of Polymeric Procyandins with Phloroglucinol: Intramolecular Rearrangements. J. Chem. Soc., Perkin Trans. 1, 1875 (1987).
111. Steenkamp, J.A., J.P. Steynberg, E.V. Brandt, D. Ferreira, and D.G. Roux: Phlobatannins: A Novel Class of Ring-isomerized Condensed Tannins. J. Chem. Soc., Chem. Commun., 1678 (1985).
112. Steynberg, J.P., D.A. Young, J.F.W. Burger, D. Ferreira, and D.G. Roux: Phlobatannins *via* Facile Ring Isomerization of Profisetinidin and Prorobinetinidin Condensed Tannin Units. J. Chem. Soc., Chem. Commun., 1013 (1986).

113. MALAN, J.C.S., J.A. STEENKAMP, J.P. STEYNBERG, D.A. YOUNG, E.V. BRANDT, and D. FERREIRA: Oligomeric Flavanoids. Part 8. The First Profisetinidins and Proguibourtinidins Based on 8-C Substituted (−)-Fisetinidol Units and Related C-ring Isomerized Analogues. J. Chem. Soc., Perkin Trans. 1, 209 (1990).

114. STEYNBERG, J.P., J.A. STEENKAMP, J.F.W. BURGER, A.D. YOUNG, and D. FERREIRA: Oligomeric Flavanoids. Part 11. Structure and Synthesis of the First Phlobatannins Related to (4α,6:4α,8)-bis-(−)-Fisetinidol-catechin Profisetinidin Triflavanoids. J. Chem. Soc., Perkin Trans. 1, 235 (1990).

115. STEYNBERG, J.P., J.F.W. BURGER, A. CRONJÉ, S.L. BONNET, J.C.S. MALAN, D.A. YOUNG, and D. FERREIRA: Oligomeric Flavanoids. Part 12. Structure and Synthesis of Phlobatannins Related to (−)-Fisetinidol-(−)-epicatechin Profisetinidins. Phytochemistry, **29**, 2979 (1990).

116. BONNET, S.L., J.P. STEYNBERG, B.C.B. BEZUIDENHOUDT, C.M. SAUNDERS, and D. FERREIRA: Oligomeric Flavanoids. Part 20. Structure and Synthesis of Phlobatannins Related to the (4β,6:4α,8)-Bis-fisetinidol-catechin Profisetinidin Triflavanoid. Phytochemistry, **43**, 215 (1996).

117. BONNET, S.L., J.P. STEYNBERG, B.C.B. BEZUIDENHOUDT, C.M. SAUNDERS, and D. FERREIRA: Oligomeric Flavanoids. Part 21. Structure and Synthesis of Phlobatannins Related to the (4β,6:4β,8)-Bis-fisetinidol-catechin Profisetinidin Triflavanoid. Phytochemistry, **43**, 229 (1996).

118. BONNET, S.L., J.P. STEYNBERG, B.C.B. BEZUIDENHOUDT, C.M. SAUNDERS, and D. FERREIRA: Oligomeric Flavanoids. Part 22. Structure and Synthesis of Phlobatannins Related to the (4α,6:4β,8)-Bis-fisetinidol-catechin Profisetinidin Triflavanoid. Phytochemistry, **43**, 241 (1996).

119. BONNET, S.L., J.P. STEYNBERG, B.C.B. BEZUIDENHOUDT, C.M. SAUNDERS, and D. FERREIRA: Oligomeric Flavanoids. Part 23. Structure and Synthesis of Phlobatannins Related to Bis-fisetinidol-epicatechin Profisetinidin Triflavanoid. Phytochemistry, **43**, 253 (1996).

120. STEYNBERG, J.P., J.F.W. BURGER, D.A. YOUNG, E.V. BRANDT, J.A. STEENKAMP, D. FERREIRA: Oligomeric Flavanoids. Part 3. Structure and Synthesis of Phlobatannins Related to (−)-Fisetinidol-(4α,6)- and (4α,8)-catechin Profisetinidins. J. Chem. Soc., Perkin Trans. 1, 3323 (1988).

121. STEYNBERG, J.P., J.F.W. BURGER, D.A. YOUNG, E.V. BRANDT, and D. FERREIRA: Oligomeric Flavanoids. Part 6. Evidence Supporting the Inversion of Absolute Configuration at 3-C Associated with Base-catalyzed A-/B-ring Interchange of Precursors Having 2,3-*trans*-3,4-*cis*-flavan-3-ol Constituent Units. Heterocycles, **28**, 923 (1989).

122. STEYNBERG, J.P., B.C.B. BEZUIDENHOUDT, J.F.W. BURGER, D.A. YOUNG, and D. FERREIRA: Novel Base-catalyzed Pyran Rearrangements of Procyanidins. J. Chem. Soc., Perkin Trans. 1, 203 (1990).

123. SAUNDERS, C.M., S.L. BONNET, J.P. STEYNBERG, and D. FERREIRA: Oligomeric Flavanoids Part 24. Controlled Biomimetic Synthesis of Profisetinidin Triflavanoid Related Phlobatannins. Tetrahedron, **52**, 6003 (1996).

124. OHARA, S., and R.W. HEMINGWAY: Condensed Tannins: The Formation of a Diarylpropan-2-ol Catechinic Acid Dimer from Base-catalyzed Reactions of (+)-Catechin. J. Wood Chem. Technol., **11**, 195 (1991).

125. PIZZI, A., E. OROVAN, and F.W. CAMERON: Cold-set Tannin-resorcinol-formaldehyde Adhesives of Lower Resorcinol Content. Holzforschung, **46**, 67 (1988).

126. MCGRAW, G.W., J.P. STEYNBERG, and R.W. HEMINGWAY: Condensed Tannins: A Novel Rearrangement of Procyanidins and Prodelphinidins in Thiolytic Cleavage. Tetrahedron Lett., **34**, 987 (1993).

127. Steynberg, P.J., J.P. Steynberg, R.W. Hemingway, D. Ferreira, and G.W. McGraw: Acid-catalyzed Rearrangements of Flavan-4-phloroglucinol Derivatives to Novel 6-Hydroxyphenyl-6a, 11b-dihydro-6*H*-[1]-benzofuro[2,3-*c*]chromenes and Hydroxyphenyl-3,2′-spirobidihydro[1]benzofurans. J. Chem. Soc., Perkin Trans. 1, 2395 (1997).

(Received November 18, 1998)

Constituents of *Lactarius* (Mushrooms)

W. M. Daniewski[1] and G. Vidari[2]

[1] Institute of Organic Chemistry, Polish Academy of Sciences, Warsaw, Poland

[2] Department of Organic Chemistry, University of Pavia, Pavia, Italy

Contents

1. Introduction

In the currently recognized 5-kingdom system of WHITTAKER, Fungi is a kingdom of its own, separated for instance from Plantae and Animalia (*1*). The kingdom of Fungi is vast and heterogeneous, comprising numerous microscopic species like molds, as well as the larger fungi (mushrooms). The latter are spore-producing fruit-bodies of fungi that in their vegetative phases live as mycelia. Larger fungi of the genus *Lactarius* belong to subdivision Basidiomycotina, order Agaricales, family Russulaceae. They nourish themselves by degrading organic waste products like plants and animal debris. Many are also important symbionts, forming mycorrhiza with higher plants which explains in some cases their preference for growing among certain kinds of trees (*2*). The genus is one of the largest in Agaricales and is distributed worldwide; more than 150 species are reported to grow in Europe where mixed forests are their typical habitat.

The name *Lactarius* has its origin in the fact that when the fruiting bodies are broken, they exude a milky cellular juice, namely, they lactate. This feature easily allows one to distinguish a *Lactarius* species from a congener *Russula* species or other similar mushrooms. Several morphological and biological features of *Lactarius* appeal to natural product scholars. Caps and stipes may be almost white, like in *L. vellereus*, or vividly colored, like in *L. scrobiculatus* and *L. rufus*. The flesh and/or latex of a few species is mild and edible (e.g. *Lactarius deliciosus*, *L. volemus*, *L. sanguifluus*), while most *Lactarius* taste pungent or bitter, and ingestion causes irritation to intestinal walls. In books of mycology this is usually ascribed to the effects of "acrid-resinous principles" of undetermined chemical structures. The burning sensation develops on the lips and tongue of an unskilled mycologist from within a few seconds up to a few minutes, helping him to recognize inedible and toxic species. In addition, also the color and taste of the exuded latex can vary with the species, and even within the same species can vary in time, slowly or very rapidly. These facts have a significant taxonomic relevance (*2*). For instance, the milky juice is permanently white and mild in *L. volemus*; it is white and becomes rapidly pungent in *L. vellereus*; it is very hot and changes from white to yellow in *L. scrobiculatus*; it is also white but becomes bitter and red in *L. fuliginosus*, and violet in *L. uvidus*.

It is evident that on a molecular basis the above remarkable phenomena can be ascribed to changes in the chemical contents of the juice and flesh of *Lactarius*, and each species seems to be endowed with its own chemical and enzymatic machinery. Moreover, even in harvested fruiting bodies, enzymatic reactions continue to occur and mushrooms must be considered "alive". Therefore, the correct procedures for extraction and isolation of compounds from such species are very important. At first, one should remember that mushrooms after collection in the forest can be stored only for short time. One of the authors (W.M. DANIEWSKI) observed that when large amounts of broken-up mushrooms are kept in a container heat is produced, as oxidation processes take place. When intact mushrooms are collected and immediately frozen in liquid nitrogen, any such transformation processes are inhibited. However, to extract non-polar compounds, among which are the precursors of pungent derivatives and other metabolites to be discussed later, such a low temperature is seldom necessary. Usually, it is recommended to freeze freshly collected intact fruiting bodies at −20°C and then soak them in hexane, EtOAc or CH_2Cl_2 (*3–9*) at −20°C for a few minutes. By contrast, extraction of mushrooms by soaking in solvents like acetone or alcohols may produce artifacts (*10–12*). In addition, traces of organic acids and even water may sometimes be harmful during concentration of the organic extracts.

One must remember that any damage to the flesh of the fruiting bodies and the resulting breakage of the cells triggers enzymatic transformations of precursors into several derivatives. Among them are the compounds involved in the changes of color and taste observed for many *Lactarius* species. To monitor such a complex cascade of reactions and to isolate the metabolites thus formed, fruiting bodies are minced without adding any solvent and different samples of the mush are then extracted with one of the above solvents at room temperature, at different times after injury. It is striking that enzymes such as some oxidases are not completely deactivated even in 30% ethanol at 25°C and extraction with this solvent allowed isolation of large amounts of variously oxygenated compounds of different skeletons. To add more difficulties, often, secondary metabolites of *Lactarius* decompose during chromatographic separations or on storage, as it will be reported later in this paper. In several instances, therefore, it may be really difficult to recognize whether an isolated compound is a true metabolite or a chemical artifact, and it is highly possible that several compounds reported as *Lactarius* metabolites in the literature, are actually artifacts. However, with respect to this issue, no critical revision of the literature has been attempted, except when compelling evidence has been accumulated.

Before this review was written four excellent reviews on fungal metabolites appeared in the literature (*13–16*). The last review comprised the literature concerning constituents of *Lactarius* up until the end of 1993.

In this review, we will discuss occurrence, chemistry, total synthesis and some biological aspects of those constituents that seem more peculiar to *Lactarius* than to other mushrooms and therefore possess taxonomic relevance. By contrast, other important constituents such as triterpenoids, sterols, polyisoprenoids, fatty acids, amino acid derivatives, etc. widely distributed in species of different genera of fungi will not be considered. The review covers literature included in Chemical Abstracts until the end of 1997. Some references, which appeared in early 1998 are also reported. The division of constituents of *Lactarius* into groups resulting from their biogenesis and chemical character used in the last review (*16*) in our opinion is very useful and has therefore been adopted in this review. In the tables of parts 1–24, structures, molecular formulas, melting points, specific rotations, sources and references of the compounds isolated are reported. Important synthetic derivatives are included, but their list is not exhaustive. Each table is preceded by the basic structure of compounds (numbered skeletons) included in the table. In cases when there is more than one basic structure, the numbers of compounds referred in the table are shown under the structures, and the compounds belonging to each structure are arranged according to their molecular formulas, the order being the same as in Chemical Abstracts. Structures represent the absolute configurations of compounds; when not established by spectroscopic methods or chemical correlations, they are inferred from biosynthetic considerations. Before each table some general remarks concerning the compounds included are presented. Also important spectroscopic features are mentioned. ^{1}HNMR data (δ in ppm, J in Hz) refer to $CDCl_3$ solutions, unless otherwise indicated.

The chemistry of the sesquiterpenes will be discussed in a separated chapter, which is divided into two parts:

a) Interconversions and reactions (part 25)

b) Total syntheses (part 26)

2. Sesquiterpenes Isolated from *Lactarius*

Sesquiterpenes of several types are the characteristic metabolites isolated from most *Lactarius* mushrooms. However, other metabolites such as alkaloids, phenols and derivatives have been found in some species and they are grouped in a separate chapter with other compounds.

The biogenetic pattern of all *Lactarius* sesquiterpenes is presented in Scheme 1 which also includes the related humulanes and sterpuranes so far not isolated from *Lactarius* species. They have been divided into classes according to their biosynthetic origin from farnesyl pyrophosphate. Only a small class of farnesane sesquiterpenes possess the skeleton of the acyclic precursor farnesol, while drimanes, guaianes and other classes arise by different farnesyl pyrophosphate cyclizations, the mode of cyclization being indicated by the numbers above the arrows. Two different cyclizations of the humulene precursor give rise to the classes of caryophyllanes and protoilludanes with the sesquiterpenes formally derived from a protoilludane precursor constituting the largest group of *Lactarius* sesquiterpenes. Cyclobutane ring contraction of the protoilludane cation may give rise to the marasmane skeleton, whereas further rearrangements of marasmanes lead to the glutinopallane, lactarane and secolactarane skeletons. In principle, the secolactarane skeleton may be formed by bond cleavage of a lactarane precursor; however, the results of some biomimetic-like reactions *in vitro* (*vide infra*) seem to indicate their origin from marasmanes. As an alternative to the protoilludane-marasmane pathway, isolactaranes may originate from rearrangement of a suitable sterpurane intermediate, even if this route in the *Lactarius* species has not been corroborated by isolation of any sterpurane sesquiterpene. Contraction of the seven-membered ring of lactaranes with loss of the C-8 carbon atom may produce the 8-nor lactarane skeleton (**15.1**), whereas loss of the C-13 carbon of marasmanes leads to the 13-normarasmane skeleton (**6.12**, **6.13**). The absolute configurations assigned to most sesquiterpenes isolated from *Lactarius* and the results of a few biosynthetic investigations (*7*, *17*, *18*) are consistent with this general scheme. Moreover, the occurrence of sesquiterpenes with different skeletons in the same species, for instance marasmane, normarasmane, isolactarane, lactarane, and secolactarane sesquiterpenes in *L. vellereus*, points to their common biogenesis.

Sesquiterpenes with drimane, farnesane, glutinopallane, protoilludane, isolactarane, and guaiane skeletons have been isolated so far from a few *Lactarius* species; therefore, they may be considered as chemotaxonomic markers. By contrast, large amounts of marasmanes,

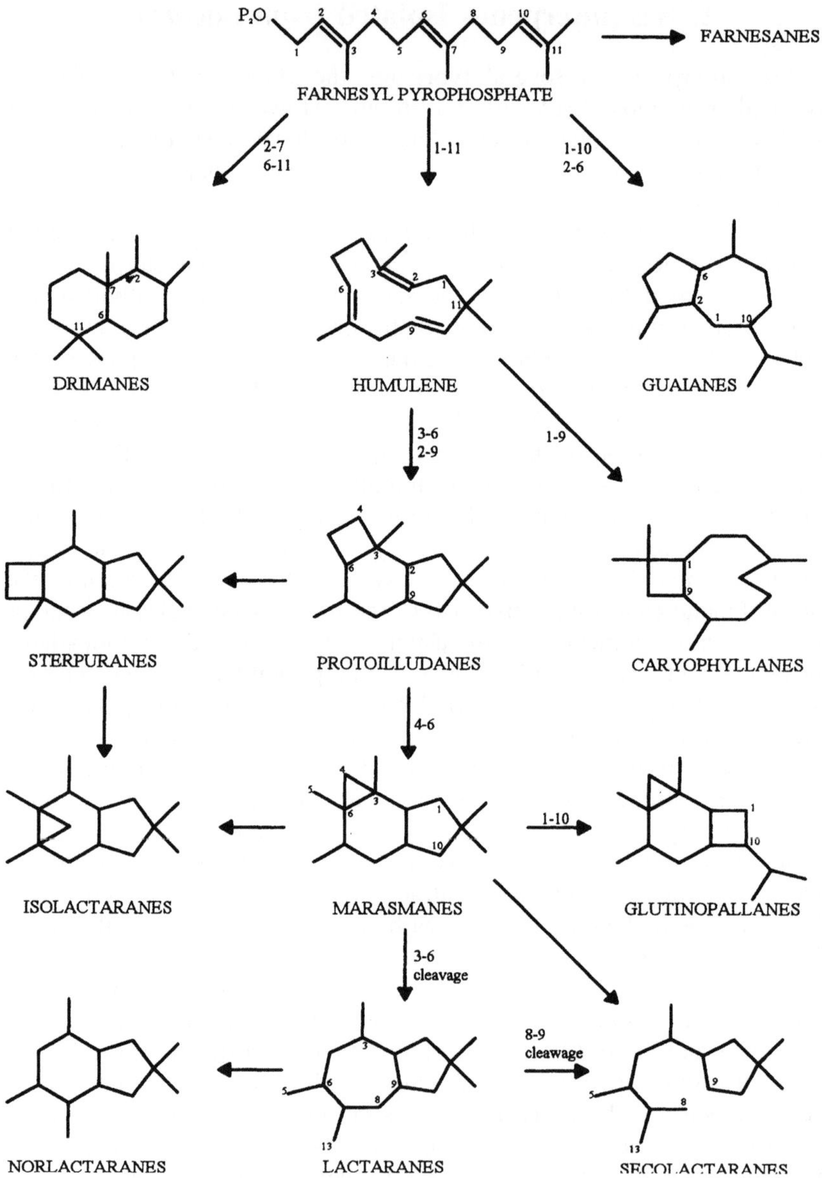

Scheme 1. Proposed biogenesis of *Lactarius* sesquiterpenes

lactaranes, and secolactaranes occur in almost all Sections (*2*) as reported in Parts 6–12, 14–16, and 18–19.

Carbons 5 and 13 of the skeletons of many marasmane, lactarane and secolactarane sesquiterpenes are linked by an oxygen atom, thus forming an extra ring, either a furan or a γ-lactone ring. In the latter the carbonyl group may be located either at C-5 or at C-13. Therefore, it was convenient to subdivide these classes of sesquiterpenes into the following groups: simple marasmane and lactarane sesquiterpenes (Parts 6 and 10, respectively), heterocyclic marasmanes (Part 7), 5-lactaranolides (Parts 11–12), 8,9-seco-5-lactaranolides (Part 14), 13-lactaranolides (Parts 16 and 17), furanolactaranes (Part 18), and 8,9-secofuranolactaranes (Part 19). Compounds with rearranged structures obtained by chemical reactions, are reported in Parts 13 and 20.

Drimane, guaiane, farnesane and caryophyllane sesquiterpenoids are typical products of plant metabolism. By contrast, sesquiterpenes with the skeletons derived from a protoilludane precursor have been isolated so far only from Basidiomycetes, but they are not unique to *Lactarius* species. In fact, marasmanes have been found, for example, also in species of the genera *Russula*, *Lentinellus*, *Auriscalpium*, *Bondarzewia*, *Vararia*, *Dichostereum*, *Peniophora*, *Artomyces*, *Marasmius*, and *Fomitopsis*; protoilludanes have been isolated from *Russula*, *Lentinellus*, *Fomitopsis*, *Clitocybe*, *Laurilia* and *Armillaria* species; isolactaranes have been isolated from *Stereum* and *Merulius* species, while lactaranes and secolactaranes are also present in *Russula*, *Lentinellus*, and *Fomitopsis* species. In fact, anatomical characteristics point to the possibility that several of these genera may form a natural group together with the genera *Lactarius* and *Russula* (Russulaceae).

Identification of some species is a difficult task, appearing frequently in section Albati, and has caused contradictory reports on the presence of certain metabolites, *e.g.* isovelleral (**6.2**), vellerolactone (**11.6**), velleral (**10.6**). We have not attempted to revise the literature in this respect and report sources of compounds as published.

Part 1. Farnesane Sesquiterpenes

A small group of farnesane sesquiterpenes, **1.1–1.9**, was isolated from *Lactarius porninsis*, a typical mushroom of European larch woods and the only species belonging to Section Zonarii Quel. (*2*) to have been investigated so far (*3*). Following the general guidelines given in the introduction, intact specimens of *Lactarius porninsis* and injured mushrooms were extracted with hexane and a series of extracts were obtained at different times after injury. It appeared that the fruiting bodies of *Lactarius porninsis* originally contain a mixture of tasteless fatty acid esters **1.3–1.9** of one farnesane sesquiterpene alcohol, called porninsol (**1.2**), and that these esters are slowly (within minutes) converted enzymatically to the aldehyde **1.1** and the free alcohol **1.2** when the fruiting bodies are damaged. Porninsal (**1.1**) and porninsol (**1.2**) both have orange-like flavors, but both also have an unpleasant though not pungent taste. This kind of enzymatic transformation occurring in *L. porninsis* is thus similar to that observed for most *Lactarius* species (*vide infra*). They originally contain tasteless long chain esters of alcohols of different types (mainly sesquiterpenes) which, when mushroom tissues are injured, are liberated in the free form and further transformed into other products which possess various biological activities and tastes. The compounds involved in these biochemical processes may thus vary from species to species; however, in general they are believed to constitute different variants of a chemical defense system that protects *Lactarius* species against parasites and predators.

The conjugated tetraene system present in porninsol, porninsal, and the esters **1.3–1.9** impart a high thermal and photochemical liability to these compounds which readily polymerize when their solutions are taken to dryness. The reader should refer to the original paper (*3*) for a description of the especially mild conditions required for the isolation of these sesquiterpenes and for recording their spectroscopic data. The UV absorption curves of **1.1–1.9** were almost superimposable indicating the same chromophore with three maxima at λ_{max} 285, 297, and 311 nm. Very similar were also the ^{1}H- and ^{13}C NMR signals of the tetraene moiety, whose stereochemistry was established by the coupling constants and NOEDS results. Interestingly, all the conjugated double bonds of the tetraene system exist in the *s-trans* conformation. The composition of the porninsol ester mixture was established by ^{1}H NMR spectra and capillary GC and GC-MS analysis of the methyl esters obtained by transesterification. The major components were saturated esters (about 86%) and the minor components were unsaturated esters (about 14%). Esters **1.3–1.9** account for about 90% of the entire ester mixture and ester **1.9** was the most abundant of all (*ca.* 64%) (*3*).

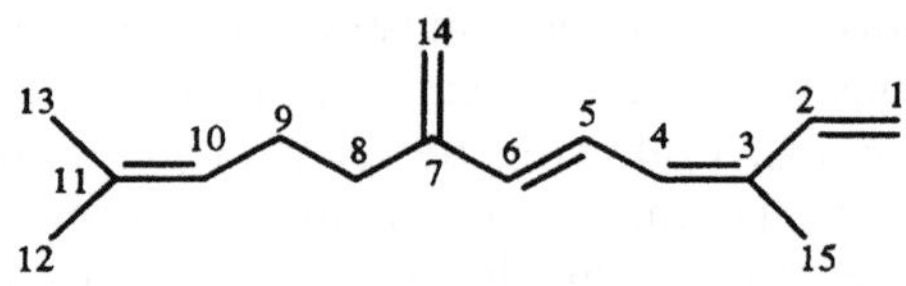

No.	Substituents, name	Formula	m.p. °C	$[\alpha]_D^{20}$	Source	Ref.
1.1	13-oxo, Porninsal	$C_{15}H_{20}O$	–	–	*L. porninsis*	(*3*)
1.2	13-OH, Porninsol	$C_{15}H_{22}O$	–	–	*L. porninsis*	(*3*)
1.3	13-miristoyloxy, Miristoylporninsol	$C_{29}H_{48}O_2$	–	–	*L. porninsis*	(*3*)
1.4	13-pentadecanoyloxy, Pentadecanoylporninsol	$C_{30}H_{50}O_2$	–	–	*L. porninsis*	(*3*)
1.5	13-palmitoleyloxy, Palmitoleylporninsol	$C_{31}H_{50}O_2$	–	–	*L. porninsis*	(*3*)
1.6	13-palmitoyloxy, Palmitoylporninsol	$C_{31}H_{52}O_2$	–	–	*L. porninsis*	(*3*)
1.7	13-linoleyloxy, Linoleylporninsol	$C_{33}H_{52}O_2$	–	–	*L. porninsis*	(*3*)
1.8	13-oleyloxy, Oleylporninsol	$C_{33}H_{54}O_2$	–	–	*L. porninsis*	(*3*)
1.9	13-stearoyloxy, Stearoylporninsol	$C_{33}H_{56}O_2$	–	–	*L. porninsis*	(*3*)

Part 2. Caryophyllane Sesquiterpenes

Alcohol **2.1** is the only example of this class, and was isolated from *Lactarius camphoratus* (*19*) which belongs to Section Olentes (*2*). The compound is presumably the product of 1–9 cyclization of a humulene precursor. The structure and absolute configuration of this new caryophyllene oxide (**2.1**) was determined by a combination of spectral data and a single-crystal X-ray analysis of the *p*-bromobenzoate derivative **2.3**. The ^{1}H NMR spectrum of **2.1** exhibited only two methyl signals, one of them at δ 1.20, together with one-proton doublet of doublets centred at δ 2.93, strongly suggesting the presence of the 4,5-epoxide. The C-12 methylene protons were shifted downfield (δ 3.84) in the 220 MHz ^{1}H NMR spectrum of the acetyl derivative (**2.2**). Moreover,

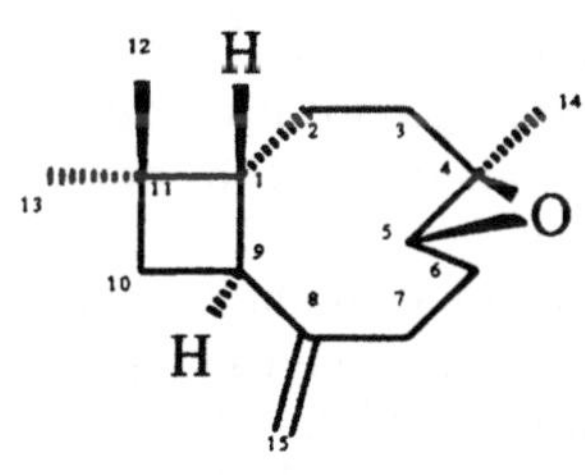

No.	Substituents, name	Formula	m.p. °C	$[\alpha]_D^{20}$	Source	Ref.
2.1	12-OH	$C_{15}H_{24}O_2$	48–53	−69.9	*L. camphoratus*	(*19*)
2.2	12-OAc	$C_{17}H_{26}O_3$	80–83	−46.4	–	(*19*)
2.3	12-OBz-*p*-Br	$C_{22}H_{27}BrO_3$	88–90	−16.3	–	(*19*)

the exocyclic methylene group is responsible for a pair of one-proton broad singlets at δ 4.90 and 5.02 ppm, respectively.

Compound **2.1** is one of the very few examples of sesquiterpenes from *Lactarius*, where one of the geminal methyl groups occurring in all skeletons has been oxidized to a hydroxymethylene function.

Part 3. Drimane Sesquiterpenes

Among the constituents of *Lactarius* mushrooms, drimane sesquiterpenes have been isolated so far only from one species of Section Uvidi (2), *i.e. Lactarius uvidus* (*20–22*). This finding is of special chemotaxonomic interest since the biosynthesis of drimane sesquiterpenes diverges completely from that of most *Lactarius* sesquiterpenes (see Scheme 1). An acetone extract of *L. uvidus* gave the known alcohol drimenol (**3.7**) and a few new representatives of the rare class of 6-oxygenated drimane sesquiterpenes, namely uvidin A (**3.5**), B (**3.6**), C (**3.9**), D (**3.10**), and E (**3.3**) (*20*), (*21*). Other interesting structural features of the uvidins are the epoxide ring at C(7)–C(8) in uvidin A, B, and C, the β-hydroxy group at C-3 in **3.6** and the α-hydroxy group at C-5 in uvidin E. Extraction of the fruiting bodies with CH_2Cl_2 at −20°C allowed isolation, in addition to the free alcohols **3.7**, **3.5**, **3.6**, and **3.10**, of a large number of low polar lipids comprising several new fatty acid esters of drimenol (**3.29**, **3.31**, **3.34**, **3.35**, **3.37**) and uvidin A (**3.28**, **3.30**, **3.32**, **3.33**, **3.36**). The most abundant were esters of 6-ketostearic acid (lactarinic acid) and (9*Z*)-$C_{18:1}$ and (9*Z*,12*Z*)-$C_{18:2}$ acids. By contrast, no ester of uvidin B, one of the major free alcohols of *L. uvidus*, could be isolated in significant amounts. The ^{1}HNMR spectra of uvidins are characterized by the signals of the four methyl groups at C-4, C-8, and C-10, which occur in the range of δ 0.81–1.50 except for the vinylic C(8)-methyl group in uvidin E (**3.3**) which is shifted downfield at δ 2.07. Particularly diagnostic for the presence of the epoxide group in uvidin A, B, C, and the corresponding esters are the signals of the C(8)-methyl group at *ca.* δ 1.4–1.5 and of the C(7)-methine at *ca.* δ 2.9–3.3; another peculiar signal of the uvidins is the singlet of the C(5)-methine, which appears in the range of δ1.9–2.2 when flanked by a carbonyl group, as in uvidin A, B, and D, while it is unexpectly shifted to high-field, below δ 1.0, in uvidin C (**3.9**) and its derivatives. For example, in 11-O-acetyluvidin C (**3.18**) the 5-H doublet is observed at δ 0.68 (*20*).

A strong positive Cotton effect at 320 nm in the CD spectra of uvidin A (**3.5**) and B (**3.6**) indicated the β-configuration of the epoxy ring, according to the anti-octant rule for αβ-epoxyketones (*20*).

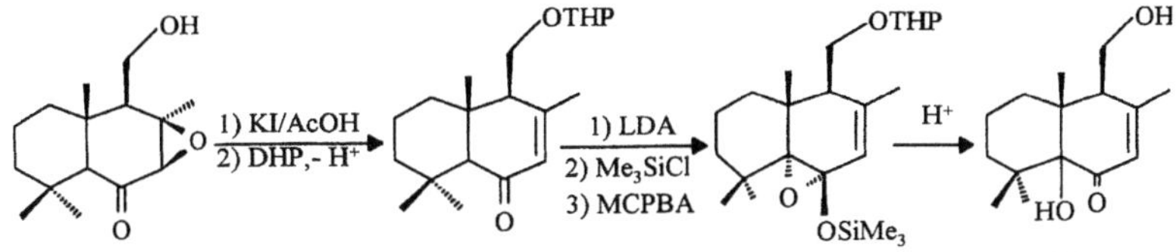

3.5 **3.22** **3.27** **3.3**

Scheme 2

Among the chemical reactions performed on uvidins in order to confirm their stereostructures, the synthesis of uvidin E (**3.3**) from uvidin A (**3.5**) (*21*) (Scheme 2) was carried out. It contained two rather interesting steps: the deepoxidation of uvidin A to 6-oxodrimenol with KI in acetic acid, and the regiospecific and stereoselective introduction of an angular α-OH group at C-5 on enone **3.22**, following the Rubottom procedure.

Exposure of uvidin A (**3.5**) or the mixture of uvidin A esters to methanolic KOH led to the new lactone **3.38**, arising by a Favorski-like rearrangement of the α,β-epoxyketone group. The same rearrangement was also observed for 11-O-ethoxyethyl uvidin A; however, in this case, the formation of the lactone ring involved the tertiary C(7)–OH group of compound **3.39**. Compounds **3.38** and **3.39** have a new sesquiterpenoid skeleton named isothapsane (*22*).

Uvidins are attractive chiral starting materials for the synthesis of highly oxidized biologically active drimane-like sesquiterpenes, as already demonstrated by the syntheses of natural (−)-cinnamodial (*23*) and (−)-cinnamosmolide (*24*) from uvidin A (**3.5**).

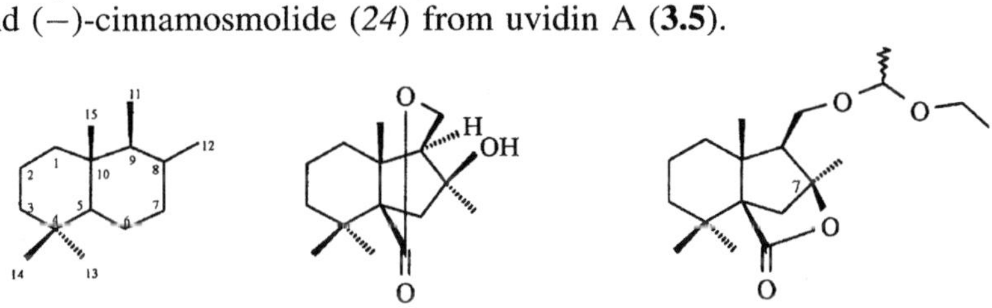

3.1 - 3.37 **3.38** **3.39**

No.	Substituents, name	Formula	m.p. °C	$[\alpha]_D^{20}$	Source	Ref.
3.1	11-OH;6-oxo7(8)-en;5α-H	$C_{15}H_{24}O_2$	84	–	–	(*20*)
3.2	3β,11-diOH;6-oxo; 7(8)-en;5α-H	$C_{15}H_{24}O_3$	222–226	+14.71	–	(*20*)
3.3	5α,11-diOH;6-oxo;7(8)-en, Uvidin E	$C_{15}H_{24}O_3$	127–129	+6.4	*L. uvidus*	(*21*)

(continued on p. 80)

(continued from p. 79)

No.	Substituents, name	Formula	m.p. °C	$[\alpha]_D^{20}$	Source	Ref.
3.4	7,11-diOH;6-oxo;7(8)-en; 5α-H	$C_{15}H_{24}O_3$	–	–	–	(21)
3.5	7β,8β-epoxy;11-OH; 6-oxo;5α-H, Uvidin A	$C_{15}H_{24}O_3$	123–124	+151.1	*L. uvidus*	(20)
3.6	7β,8β-epoxy;3β,11-diOH; 6-oxo;5α-H, Uvidin B	$C_{15}H_{24}O_4$	180–181	+171	*L. uvidus*	(20)
3.7	11-OH;7(8)-en;5α-H, Drimenol	$C_{15}H_{26}O$	97–98	−22	*L. uvidus*	(20)
3.8	11-OH;6-oxo;5α,8α-H	$C_{15}H_{26}O_2$	108–110	+34.30	–	(20)
3.9	7β,8β-epoxy;6β,11-diOH; 5α-H, Uvidin C	$C_{15}H_{26}O_3$	107–111	–	*L. uvidus*	(20, 21)
3.10	7β,11-diOH;6-oxo;5α,8α-H, Uvidin D	$C_{15}H_{26}O_3$	152–154	–	*L. uvidus*	(21)
3.11	8β,11-diOH;6-oxo;5α-H	$C_{15}H_{26}O_3$	168–169	+44.76	–	(20)
3.12	3β,11-diOH;6-oxo;5α,8α-H	$C_{15}H_{26}O_3$	207–209	+30.64	–	(20)
3.13	7α,11-diOH;6-oxo;5α,8α-H	$C_{15}H_{26}O_3$	150–152	–	–	(21)
3.14	11-OH;5α,8α-H	$C_{15}H_{28}O$	106–108	+9.2	–	(20)
3.15	5α-OH;11-OAc; 6-oxo;7(8)-en	$C_{17}H_{26}O_4$	148–149	–	–	(21)
3.16	11-OH;7-OAc;6-oxo; 7(8)-en;5α-H	$C_{17}H_{26}O_4$	–	–	–	(21)
3.17	7β,8β-epoxy;11-OAc; 6-oxo;5α-H	$C_{17}H_{26}O_4$	81.5–82.5	+155.6		(20)
3.18	7β,8β-epoxy;6β-OH; 11-OAc;5α-H	$C_{17}H_{28}O_4$	75–77	–	–	(20, 21)
3.19	3β,11-diOAc;6-oxo; 7(8)-en;5α-H	$C_{19}H_{28}O_5$	104–106	+60.78	–	(20)
3.20	7β,8β-epoxy;3β,11-diOAc; 6-oxo;5α-H	$C_{19}H_{28}O_6$	150–152	+140.2	–	(20)
3.21	7β,8β-epoxy;6β,11-diOAc; 5α-H	$C_{19}H_{30}O_5$	–	–	–	(21)
3.22	11-OTHP;6-oxo; 7(8)-en;5α-H	$C_{20}H_{32}O_3$	oil	–	–	(21)
3.23	7-OH;11-OTHP;6-oxo; 7(8)-en;5α-H	$C_{20}H_{32}O_4$	oil	–	–	(21)
3.24	11-OTHP;6-oxo;5α,8α-H	$C_{20}H_{34}O_3$	–	–	–	(21)
3.25	7-OAc;11-OTHP; 6-oxo;7(8)-en;5α-H	$C_{22}H_{34}O_5$	–	–	–	(21)
3.26	6-OSiMe$_3$;11-OTHP; 5(6),7(8)-dien	$C_{23}H_{40}O_3Si$	–	–	–	(21)
3.27	5α,6α-epoxy;6β-OSiMe$_3$; 11-OTHP;7,8-en	$C_{23}H_{40}O_4Si$	–	–	–	(21)
3.28	7β,8β-epoxy; 11-palmitoyloxy;6-oxo; 5α-H, Palmitoyluvidin A	$C_{31}H_{54}O_4$	–	–	*L. uvidus*	(22)
3.29	11-palmitoyloxy; 7(8)-en; 5α-H, Palmitoyldrimenol	$C_{31}H_{56}O_2$	–	–	*L. uvidus*	(22)

(continued on p. 81)

(continued from p. 80)

No.	Substituents, name	Formula	m.p. °C	$[\alpha]_D^{20}$	Source	Ref.
3.30	7β,8β-epoxy;11-linoleyloxy; 6-oxo;5α-H, Linoleyluvidin A	$C_{33}H_{54}O_4$	–	–	*L. uvidus*	(22)
3.31	11-linoleyloxy;7(8)-en; 5α-H, Linoleyldrimenol	$C_{33}H_{56}O_2$	–	–	*L. uvidus*	(22)
3.32	7β,8β-epoxy;11-oleyloxy; 6-oxo;5α-H, Oleyluvidin A	$C_{33}H_{56}O_4$	–	–	*L. uvidus*	(22)
3.33	7β,8β-epoxy;11-(6′-oxostearoyloxy);6-oxo;5α-H, 6′-Ketostearoyluvidin A	$C_{33}H_{56}O_5$	52–53	+101.8	*L. uvidus*	(22)
3.34	11-oleyloxy;7(8)-en;5α-H, Oleyldrimenol	$C_{33}H_{58}O_2$	–	–	*L. uvidus*	(22)
3.35	11-(6′-oxostearoyloxy); 7(8)-en;5α-H, 6′-Ketostearoyldrimenol	$C_{33}H_{58}O_3$	22–23	+7.24	*L. uvidus*	(22)
3.36	7β,8β-epoxy;11-stearoyloxy;6-oxo;5α-H, Stearoyluvidin A	$C_{33}H_{58}O_4$	–	–	*L. uvidus*	(22)
3.37	11-stearoyloxy;7(8)-en; 5α-H, Stearoyldrimenol	$C_{33}H_{60}O_2$	–	–	*L. uvidus*	(22)
3.38	(see formula)	$C_{15}H_{24}O_3$	206–208	−25.7	–	(22)
3.39	(see formula)	$C_{19}H_{32}O_4$	–	+40.4	–	(22)

Part 4. Guaiane Sesquiterpenes

Guaiane sesquiterpenes, which are not formed by the same biosynthetic pathway as the types of sesquiterpenes found in the pungent *Lactarius* species (Scheme 1), can be considered chemotaxonomic markers of species belonging to Section Dapetes. The latter mushrooms are characterized by the secretion of a strongly colored milky juice and are usually edible and of pleasant taste (*2*).

The latex of each species contains a characteristic mixture of colored sesquiterpenes responsible for the natural orange, red, green, or even blue color. The structures of a dozen of guaiane sesquiterpenes were determined by chemical and spectral methods. The pattern of the olefinic protons (δ 5.15–9.96) in the ^{1}H NMR spectra, and the UV/Vis spectra are particularly indicative of the dihydroazulene or azulene guaiane structure; the λ_{max} of the electronic spectra have been collected in an earlier review on pigments of fungi (*25*) covering the literature to the early part of 1986. Compounds listed in Part 4 are extraordinarily chemically sensitive and could be isolated by employing very mild extraction and purification conditions. For example, the two pigments of *L. indigo*, namely the wine red lactaroviolin (**4.1**) and the blue stearoyldeterrol (**4.11**), are rapidly converted to an intractable green material upon addition of MeOH to the acetone solutions, or on attempted chromatography using several solvents (*6*). Both ester **4.13** and the corresponding alcohol **4.9** polymerized in air (*4*). Similarly, delicial (**4.6**) rapidly polymerized when exposed to light (*5*). Small amounts of delicial could be obtained only by flash chromatography in the dark and with cold solvent, on silica prewashed with diethyl ether (*5*).

Most isolated guaiane sesquiterpenes have a formyl or a free hydroxymethyl group at C-4. However, recent results have proved that these compounds do not occur in undamaged fruiting bodies, but are formed enzymatically from fatty acid ester precursors, *i.e.* **4.10**, **4.11**, **4.12**, and **4.13**, in injured specimens (*5*), (*26*). Thus, the orange-yellow esters **4.10** and **4.13** (occurring in the ratio 15 and 85%, respectively) were the only sesquiterpenes isolated from *L. deterrimus* and *L. deliciosus* when young, undamaged fruiting bodies were immersed in liquid nitrogen at their growing site, and subsequently extracted with hexane at −20°C (*5*). By contrast, when fresh specimens of the two mushrooms were ground in a meat grinder to simulate an injury, and the mush was then extracted with hexane, the aldehydes **4.1** and **4.6**, the free alcohols **4.7** and **4.9**, and lactarazulene (**4.3**) were isolated (*5*). The carrot-colored latex of *L. deliciosus* and *L. deterrimus* with time assumes a green color due to the formation of violet and blue compounds (lactaroviolin (**4.1**) and deterrol (**4.7**), and their mixing with the yellow compounds (the alcohol **4.9**, the fatty acid esters of **4.9**, and delicial (**4.6**)) already present or also formed (*5*). In accordance with these findings, when young and undamaged fruiting bodies of *L. sanguifluus* were extracted with hexane in the cold, only the yellow ester **4.13** and the red ester **4.12** were detected (*26*). However, an EtOAc extract of the same mushrooms that had been ground 30 min prior to extraction, yielded the free alcohol **4.8** (*26*).

Interestingly, a few species of the section Dapetes growing in different parts of the world seem to contain different metabolites. For instance, aldehyde **4.5** has been isolated from Indian (*27*) but not from European specimens of *L. deterrimus*, while lactarofulvene (**4.2**) was isolated from Californian (*28*) but not from European specimens of *L. deliciosus* (*5*). An explanation of these apparent differences between specimens grown in different continents may be the existence of subspecies (*5*) or a change in the metabolism related to different habitats, or it may be due to the formation of artifacts during the extraction. A different pattern of aromatic compounds (not sesquiterpenes) was produced by *Lactarius deliciosus* when the fungus was grown in liquid cultures (*29*). Their structures will be reviewed later.

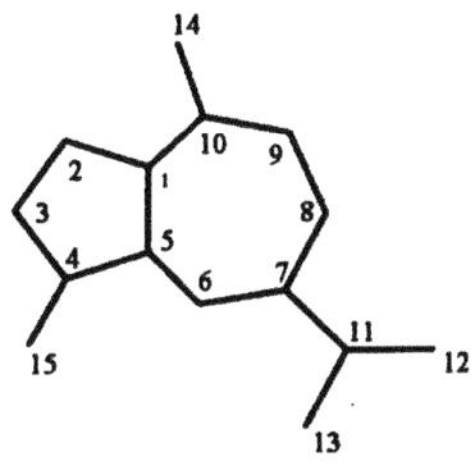

No.	Substituents, name	Formula	m.p. °C	$[\alpha]_D^{20}$	Source	Ref.
4.1	15-oxo;1(10),2(3),4(5),6(7),8(9),11(12)-esaen, Lactaroviolin	$C_{15}H_{14}O$	58	–	*L. deliciosus*, *L. deterrimus*, *L. indigo*, *L. sanguifluus*, *L. semisanguifluus*	(*4, 5, 6, 30, 31, 32*)
4.2	1(5),2(3),4(15),6(7),9(10),11(12)-esaen, Lactarofulvene	$C_{15}H_{16}$	–	–	*L. deliciosus*	(*4, 28*)
4.3	1(10),2(3),4(5),6(7),8(9),11(2)-esaen, Lactarazulene	$C_{15}H_{16}$	35.5	–	*L. deliciosus*, *L. deterrimus*, *L. semisanguifluus*	(*4, 5, 30, 33*)
4.4	15-oxo;1(2),3(4),5(6),7(11),9(10)-pentaen	$C_{15}H_{16}O$	–	–	*L. sanguifluus*	(*34*)
4.5	15-oxo;1(10),2(3),4(5),6(7),8(9)-pentaen	$C_{15}H_{16}O$	59–60	–	*L. deterrimus*; *L. sanguifluus*	(*27, 34*)
4.6	15-oxo;1(2),3(4),5(6),9(10),11(12)-pentaen, Delicial	$C_{15}H_{16}O$	oil	–	*L. deliciosus*, *L. deterrrimus*	(*5*)
4.7	15-OH; 1(10),2(3),4(5),6(7),8(9),11(12)-esaen, Deterrol	$C_{15}H_{16}O$	100–101	–	*L. deliciosus*, *L. deterrimus*	(*5*)
4.8	15-OH;1(2),3(4),5(6),7(11),9(10)-pentaen, Sangol	$C_{15}H_{18}O$	oil	–	*L. sanguifluus*	(*26*)
4.9	15-OH;1(2),3(4),5(6),9(10),11(12)-pentaen	$C_{15}H_{18}O$	–	–	*L. deliciosus*, *L. deterrimus*, *L. salmonicolor*, *L. sanguifluus*, *L. semisanguifluus*	(*4, 5, 30*)
4.10	15-linoleyloxy;1(2),3(4),5(6),9(10),11(12)-pentaen	$C_{33}H_{48}O_2$	–	–	*L. deterrimus*, *L. deliciosus*	(*5*)

(continued on p. 84)

(continued from p. 83)

No.	Substituents, name	Formula	m.p. °C	$[\alpha]_D{}^{20}$	Source	Ref.
4.11	15-stearoyloxy;1(10),2(3), 4(5),6(7),8(9),11(12)-esaen, Stearoyldeterrol	$C_{33}H_{50}O_2$	–	–	*L. indigo*	(*6*)
4.12	15-stearoyloxy;1(2),3(4), 5(6),7(11),9(10)-pentaen, Stearoylsangol	$C_{33}H_{52}O_2$	–	–	*L. sanguifluus*	(*26*)
4.13	15-stearoyloxy;1(2),3(4), 5(6),9(10),11(12)-pentaen	$C_{33}H_{52}O_2$	–	–	*L. deliciosus*, *L. salmonicolor*, *L. deterrimus* *L. sanguifluus*, *L. semisanguifluus*	(*4, 5, 26, 30*)
4.14	Dimers	?				(*5, 6*)

Part 5. Protoilludane Sesquiterpenes

Violascensol (**5.1**) and the corresponding 6-ketostearoyl ester **5.2** are the only protoilludane sesquiterpenes isolated so far from a *Lactarius* species (*35*). This finding was of special interest since it demonstrated that biosynthetic routes leading to the protoilludane skeleton can be present in *Lactarius* cells and gave credit to the mode of biosynthesis suggested for the majority of *Lactarius* sesquiterpenes (Scheme 1). It is worth noting that 6-ketostearic acid (also named lactarinic acid) is a characteristic fatty acid of *Lactarius* mushrooms, where it usually occurs not in a free form but combined in the form of esters of various sesquiterpene alcohols (*see infra*). A synthesis of lactarinic acid is described in reference (*35*). The structures **5.1** and **5.2**, particularly the position of the oxygenated group at C-15 and the relative configuration of the molecules were determined by 2D-NMR spectra and NOESY experiments. The UV absorption at 264 nm was diagnostic for the ketone carbonyl group at C-5 conjugated with the tetrasubstituted double bond between C-6 and C-7. The isolated methylene group at C-4, flanking the carbonyl group, gave rise to a characteristic AB quartet at δ 2.62 in **5.1** and at δ 2.68 in **5.2**.

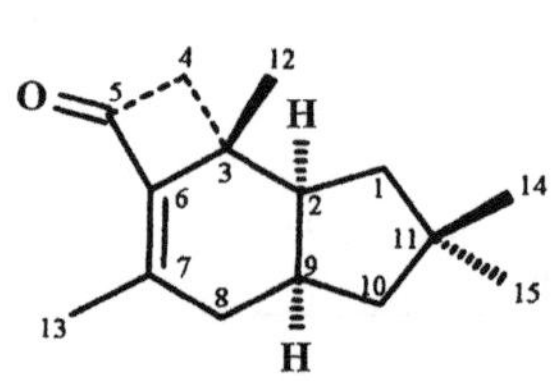

No.	Substituents, name	Formula	m.p. °C	$[\alpha]_D^{20}$	Source	Ref.
5.1	15-OH, Violascensol	$C_{15}H_{22}O_2$	–	−105.4	*L. violascens*	*(35)*
5.2	15-(6′-oxostearoyloxy), 6′-Ketostearoylviolascensol	$C_{33}H_{54}O_4$	–	−35.6	*L. violascens*	*(35)*

3. Introduction to Parts 6 and 7 – Velutinal Esters and Related Sesquiterpenes

Marasmane sesquiterpenes, possibly resulting from contraction of the cyclobutane ring of the protoilludane skeleton (Scheme 1), can be divided into two groups: tricyclic (Part 6) and heterocyclic (tetracyclic) (Part 7). The tricyclic constituents were divided into three subgroups: simple marasmanes (**6.1**–**6.8**), isomarasmanes (**6.9**–**6.11**) and 13-normarasmanes (**6.12**–**6.22**).

Among all constituents of *Lactarius* species, velutinal esters (**7.28** and **7.30**) deserve special consideration. Stearoylvelutinal (**7.30**) was originally isolated by a French group from *Lactarius velutinus* (*36*), during the search for the substances responsible for the intense blue colour of the cystidia or lacticifers of several *Lactarius* touched with the "sulpho-vanillin mixture". This reagent is used in systematic mycology for identification purposes (*8*). Independently, almost at the same time, Swedish authors isolated stearoyl- (**7.30**) and 6′-ketostearoylvelutinal (**7.28**) from *L. vellereus* and *L. necator* (*37*), in an attempt to clarify the formation of artifacts during extraction and isolation of fungal metabolites. Since then, most *Lactarius* species have been shown to contain velutinal esters, even if several important exceptions are known (*vide infra*). Each ester may occur alone or in a mixture. For example, *L. piperatus*, *L. bertillonii*, and *L. vellereus* contain ester **7.30** alone (*38*), *L. chrysorrheus* (*9*) contains **7.28**, while *L. rufus*, *L. necator*, and *L. trivialis* contain approximately 90% **7.28** and 10% **7.30** (*38*). Traces of other not yet identified esters of velutinal have also been detected in some species (*8*, *36*). It is remarkable that velutinal esters are not unique to *Lactarius* species and have been found in a number of other genera, for example in *Russula*, *Lentinellus*, *Auriscalpium*, *Bondarzewia*, *Vararia*, *Peniophora*, and *Artomyces* (*8*, *36*, *38*, *39*).

Velutinal esters are the only observable sesquiterpenes occurring in undamaged fruiting bodies, when worked up under carefully controlled conditions (see introductory remarks). If extraction is not carried out properly, they are rapidly degraded and artifacts are formed (*vide infra*).

These colorless and tasteless esters are contained as an emulsion (the latex) in specialised hyphae of *Lactarius* (*8*, *40*) and are apparently biologically inactive (*7*). A physical injury to mushroom tissues by man, parasite or any predator (*41*) triggers a complex cascade of enzymatic reactions of velutinal esters that in each mushroom gives rise to a characteristic pattern of sesquiterpenes. Anyone may perceive these transformations by noting the changes of colour and/or taste of the latex and flesh of damaged fruiting bodies. Novel marasmane (Parts 6, 7), lactarane (Parts 10, 11, 15, 16, 18), secolactarane (Parts 14, 19), and possibly isolactarane sesquiterpenes (Part 9) are thus formed; in these compounds carbons 5 and 13 usually exhibit a 1,4-dialdehyde, a hydroxyaldehyde or a diol functionality, or form a lactone or a furan ring. The time required for each reaction to occur may vary from a few seconds to several hours or more and depends on the species. For example, significant amounts of pungent isovelleral (**6.2**) (*7*) and chrysorrhedial (**10.2**) (*9*) appear in hexane extracts of *L. vellereus* and *L. scrobiculatus*, respectively, made less than one minute after injury. In damaged fruiting bodies of *L. bertillonii*, the equally pungent dialdehyde velleral (**10.6**) is formed within seconds and is completely converted to vellerol (**10.9**) within minutes, while vellerolactone (**11.6**), which is already present in extracts made 5 min after injury, becomes the major extractable sesquiterpenoid component (*ca.* 75%) 30 min after injury (*42*). By contrast, vellerdiol (**10.14**), which is produced by enzymatic reduction of vellerol (**10.9**), appears in extracts of *L. vellereus* only several hours after grinding at 22° (*7*). In conclusion, extraction of fruiting bodies at different times after an injury usually affords a different pattern of metabolites and may suggest possible biosynthetic pathways (*9*). Most details of these enzymatic transformations are still unknown; however, 1,4-unsaturated dialdehydes appear to be formed immediately from the velutinal esters, while most of the other sesquiterpenes are subsequently produced by enzymatic transformations of the carbonyl groups of the aldehydes. In this context, the isolation of vellerolactone (**11.6**) together with velleral (**10.6**) and vellerol (**10.9**) from *L. bertillonii* (*42*), isovelleral (**6.2**), isovellerol (**7.11**), and lactone **7.3** from *L. vellereus* (*7*), (*43*), blennin A (**11.26**), piperdial (**10.11**), and piperalol (**10.16**) from *L. torminosus* (*44*, *45*, *46*), 7-*epi*-piperdial (**10.12**), *epi*-piperalol (**10.17**), 7-*epi*-pipertriol (**10.20**), and lactarorufin N (**11.22**) in *L. necator* (*47*, *17*, *48*), chrysorrhedial (**10.2**), chrysorrheal (**10.10**), chrysorrhelactone (**16.1**), and lactaroscrobiculide A (**16.2**) in *L. srobiculatus* (*9*), (*49*), (*50*), suggest a pattern.

It is worth noting that *Lactarius* sesquiterpenes derived from velutinal esters (**7.28**, **7.30**) have the same configuration of the related antibiotic marasmic acid (*51*), which is opposite to that of hirsutic acid, another fungal metabolite (*52*) biosynthetically derived from a humulene precursor.

The ^{1}H NMR spectra of marasmane sesquiterpenes possess very characteristic signals which allow their identification. The geminal cyclopropane protons at C-4 give rise to a pair of doublets with $J=$ 4–5 Hz. The chemical shifts of these protons depend upon other substituents present in the molecule. When no carbonyl or ester group is at C-5 or at C-12, the protons resonate at high fields ranging from δ 0.25 to 1.0. However, in case of marasmane aldehydes (e.g. isovelleral **6.2**), lactones (e.g. **7.3**), and rubrocintal derivatives (Part 7), the shifts of C-4 protons range from δ 0.90 to 2.0. The geminal methyl groups at C-11 appear as singlets at *ca.* δ 0.9–1.1, while the C-12 methyl (where applicable) resonates at lower field (*ca.* δ 1.15–1.5). In addition, the ^{1}H NMR spectra of lactones and lactols are characterized by the presence of an ABq system (J_{AB}=10–13 Hz) for the methylene protons at C-13 which, in 7(8)-ene derivatives, exhibit an additional allylic coupling with H-8 and a homoallylic coupling with H-9; in free hemiacetals, two singlets at δ 5.2–5.9 signal the presence of two epimers at C-5. In the corresponding esters the signal of H-5 is shifted to *ca.* δ 6.2, as expected.

Part 6. Marasmane, Isomarasmane, and Normarasmane Sesquiterpenes

The hot tasting crystalline isovelleral (**6.2**) was isolated, together with velleral (**10.6**), from *L. vellereus* by P.H. List (*53*) in 1969; however, only its partial structure was reported at that time. Later, during the pioneering work of Swedish authors on the chemical constituents of *Lactarius* species they reported the complete structure of isovelleral with full spectroscopic data. It was the first marasmane sesquiterpene isolated from *Lactarius* species (*54*), (*55*). Later, isovelleral was discovered, either alone or together with velleral, in other pungent species, where it is rapidly (within seconds) formed from velutinal ester precursors after an injury to fruiting bodies (see above). Given its manner of formation and its striking biological activities (*vide infra*), isovelleral (**6.2**) is therefore considered to be part of a chemical defense system of the mushrooms.

Of the other naturally occurring marasmanes included in Part 6, the very stable 5,10α,13-trihydroxymarasm-7(8)-ene (**6.6**) and lactaropallidine (**6.5**) (*56*) are of special interest. Triol **6.6** is one of the few sesquiterpenes of *Lactarius* exhibiting an oxygenated group on one of the methylene carbons of the cyclopentane ring; lactaropallidine was assigned its absolute configuration by CD measurement and was chemically correlated with stearoylvelutinal (**7.30**) (*56*) (see chemical part), thus indicating the absolute configuration of all sesquiterpenes derived from velutinal esters. The enantioselective synthesis of isovelleral (**6.2**) (*57*) definitely confirmed this assignment.

Part 6 also includes a few isomarasmanes (**6.9**–**6.11**) and normarasmanes (**6.12**–**6.22**). The former are synthetic compounds and were obtained by thermal isomerization of the corresponding marasmanes (see chemical part). The 13-normarasmane isomers **6.12** and **6.13** are among the few known norsesquiterpenes of *Lactarius*. The two compounds were isolated (*58*) together with compound **6.6** from an ethanol extract of *Lactarius vellereus*. The normarasmane ketones may derive in mushrooms from lactaropallidine (**6.5**) by β-deformylation and oxidation at C-7.

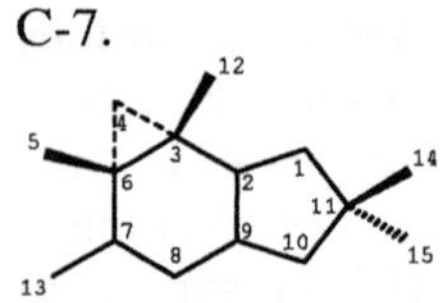

6.1-6.8

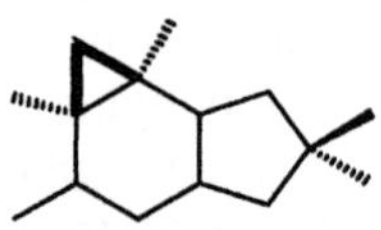

6.9-6.11

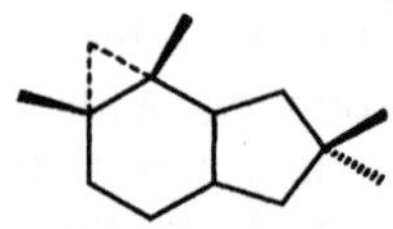

6.12-6.22

No.	Substituents, name	Formula	m.p. °C	$[\alpha]_D^{20}$	Source	Ref.
6.1	12-$^2H_3$;5,13-dioxo; 7(8)-en;2α,9α-H, [12-$^2H_3$]-Isovelleral	$C_{15}H_{17}D_3O_2$	95–97	+237	–	(*59*)
6.2	5,13-dioxo;7(8)-en; 2α,9α-H, Isovelleral	$C_{15}H_{20}O_2$	105–106	+293	*L. vellereus*, *L. quietus*, *L. deceptivus*, *L. subvellereus*, *L. tomentoso-marginatus*, *L. rufus* *L. torminosus*, *L. flexuosus*, *L. piperatus*, *L. pergamenus*	(*3, 7, 17, 33, 40, 41, 53, 54, 60*)
6.3	9α-OH;5,13-dioxo; 7(8)-en;2α-H	$C_{15}H_{20}O_3$	oil	+92	–	(*7, 61, 62*)
6.4	5,13-diOH;7(8)-en; 2α,9α-H Isovellerdiol	$C_{15}H_{24}O_2$	110–111	–2	–	(*54*)
6.5	5,13-diOH;8-oxo;2α,7α, 9α-H, Lactaropallidine	$C_{15}H_{24}O_3$	–	–43.8	*L. pallidus*	(*56*)
6.6	5,10α,13-triOH;7(8)-en; 2α,9α-H	$C_{15}H_{24}O_3$	–	–	*L. vellereus*	(*63*)

(continued on p. 89)

(continued from p. 88)

No.	Substituents, name	Formula	m.p. °C	$[\alpha]_D^{20}$	Source	Ref.
6.7	5,7α,8α,13-tetraOH; 2α,9α-H	$C_{15}H_{26}O_4$	158.5–159.5	−73	–	*(54)*
6.8	5,10α,13-triOAc; 7(8)-en;2α,9α-H	$C_{21}H_{30}O_6$	75–76	+7.0	–	*(63)*
6.9	12-$^2H_3$;5,13-dioxo; 7(8)-en;2α,9α-H, [12-$^2H_3$]-Isoisovelleral	$C_{15}H_{17}D_3O_2$	67–69	−88.3	–	*(59)*
6.10	5,13-dioxo;7(8)-en; 2α,9α-H Isoisovelleral	$C_{15}H_{20}O_2$	72.5–74.5	−63	–	*(57, 59)*
6.11	9α-OH;5,13-dioxo; 7(8)-en;2α-H	$C_{15}H_{20}O_3$	–		–	*(61)*
6.12	5,7α-diOH;8-oxo; 2α,9α-H	$C_{14}H_{22}O_3$	88–91	−125.9	*L. vellereus*	*(58, 64)*
6.13	5,8α-diOH;7-oxo; 2α,9α-H	$C_{14}H_{22}O_3$	oil	−83.3	*L. vellereus*	*(58, 64)*
6.14	5,7α-diOAc; 8-oxo;2α,9α-H	$C_{18}H_{26}O_5$	oil	−76.4	–	*(58)*
6.15	5,8α-diOAc;7-oxo; 2α,9α-H	$C_{18}H_{26}O_5$	oil	−22.3	–	*(58)*
6.16	7α-OH;5,8α-diOAc; 2α,9α-H	$C_{18}H_{28}O_5$	–	–	–	*(58)*
6.17	7α-OH;5,8β-diOAc; 2α,9α-H	$C_{18}H_{28}O_5$	–	–	–	*(58)*
6.18	8α-OH;5,7α-diOAc; 2α,9α-H	$C_{18}H_{28}O_5$	–	–	–	*(58)*
6.19	8β-OH;5,7α-diOAc; 2α,9α-H	$C_{18}H_{28}O_5$	–	–	–	*(58)*
6.20	5,7α,8α-triOAc;2α,9α-H	$C_{20}H_{30}O_6$	–	–	–	*(58)*
6.21	5,7α,8β-triOAc;2α,9α-H	$C_{20}H_{30}O_6$	–	–	–	*(58)*
6.22	5,7β,8α-triOAc;2α,9α-H	$C_{20}H_{30}O_6$	–	–	–	*(58)*

Part 7. Heterocyclic Marasmane Sesquiterpenes

The group of heterocyclic marasmane sesquiterpenes includes hemiacetals velutinal (**7.12**), isovellerol (**7.11**), rubrocinctal A (**7.4**), and their derivatives and lactones which are unsubstituted or possess up to three hydroxy groups. Velutinal esters **7.28** and **7.30** have already been illustrated. The free hemiacetal velutinal (**7.12**) does not occur in intact fruiting bodies where, instead, is contained in the form of esters (see above); however, small amounts of **7.12** were detected by tlc in a hexane extract of injured *L. vellereus* within the first minutes after grinding at 4° (*7*). Moreover, velutinal could be obtained by ethanolysis at 25° of velutinal esters in 1 mM NaOEt/EtOH and rapid chromatography on prewashed SiO_2 or Al_2O_3 (*37*).

Isovellerol (**7.11**) was first discovered in 1971 by SHIGEO NOZOE and collaborators (*65*) from extracts of the cultured mycelium of *Fomitopsis insularis*. Later, it was again isolated from a hexane extract of injured fruiting bodies of *L. vellereus*, where it is enzymatically formed from isovelleral (**6.2**) (*7*). The same mushroom gave a group of marasmane lactones (**7.3**, **7.5–7.9**, **7.13–7.15**) when fruiting bodies were left in EtOH for two months (*43*), (*66*), (*58*), (*67*). This finding is amazing in view of the fact that stearoylvelutinal (**7.30**), considered to be the only sesquiterpene occurring in significant amounts in uninjured *L. vellereus* (*7*), is rapidly degraded in alcohols to afford lactarane or secolactarane derivatives (*vide infra*) (*68*), (*12*), but no marasmanes. Evidently, some lipophylic microenvironments present in the mushrooms protect **7.30** from prolonged contact with EtOH and allow extensive oxidation of substrates. In fact, the unsubstituted lactone **7.3** may be formed by oxidation of hemiacetal **7.11** and the other hydroxylated lactones by further oxidations of compound **7.3**. In lactone **7.14** H-8 and H-9 have the unusual *cis* configuration, as established unequivocally by the value of the coupling constant ($J = O$). In the case of stereoisomer **7.13**, $J_{8,9}$ is equal to 11.6 Hz, showing *trans* configuration. In analogy with isovellerdiol (**6.4**) (*54*), the 7,8-double bond of **7.3** was dihydroxylated with OsO_4, from the convex side of the molecule and gave the corresponding 7α,8α-diol (**7.13**) (*43*). X-ray analysis of compound **7.13** and a derivative (**7.19**) of lactone **7.7** definitely confirmed the structures.

The aldehyde rubrocinctal A (**7.4**), the 12-carboxymethyl ester rubrocinctal B (**7.16**), and the corresponding 6-oxostearoyl esters **7.27** and **7.31** were isolated from *Lactarius rubrocinctus* Section Ichorati (*2*), (*69*). They are of special interest because they are, along with glutinopallal esters **8.2** and **8.3**, the only examples of 12-oxygenated sesquiterpenes isolated from *Lactarius* species.

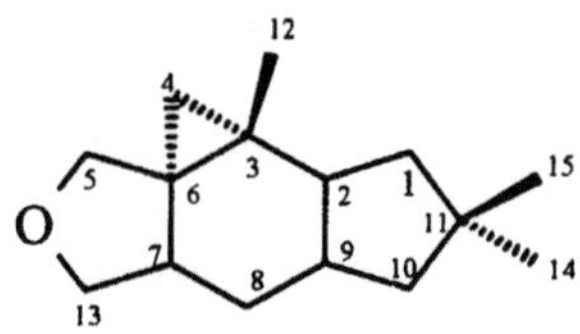

No.	Substituents, name	Formula	m.p. °C	$[\alpha]_D{}^{20}$	Source	Ref.
7.1	12-$^2H_3$;5-OH;7(8)-en; 2α,9α-H, [12-$^2H_3$]-Isovellerol	$C_{15}H_{19}D_3O_2$	–	–	–	(*18*)

(*continued on p. 91*)

(continued from p. 90)

No.	Substituents, name	Formula	m.p. °C	$[\alpha]_D^{20}$	Source	Ref.
7.2	12-$^2H_3$;7α,8α-epoxy; 5α-OH;2α,9α-H, [12-$^2H_3$]-Velutinal	$C_{15}H_{19}D_3O_3$	–	–	–	*(18)*
7.3	5-oxo;7(8)-en;2a,9a-H	$C_{15}H_{20}O_2$	oil	+91.3	*L. vellereus, L. controversus*	*(43, 70)*
7.4	5-OH;12-oxo;7(8)-en; 2α,9α-H, Rubrocinctal A	$C_{15}H_{20}O_3$	–		*L. rubrocinctus*	*(69)*
7.5	2α-OH;5-oxo; 7(8)-en;9αH	$C_{15}H_{20}O_3$	oil	+33.4	*L. vellereus*	*(66)*
7.6	9α-OH;5-oxo; 7(8)-en;2αH	$C_{15}H_{20}O_3$	oil	+2.9	*L. vellereus*	*(66)*
7.7	10α-OH;5-oxo; 7(8)-en;2α,9αH	$C_{15}H_{20}O_3$	oil	+61.8	*L. vellereus*	*(66)*
7.8	14-OH;5-oxo;7(8)-en; 2α,9αH	$C_{15}H_{20}O_3$	oil	+83.2	*L. vellereus*	*(66)*
7.9	9α,10α-diOH;5-oxo; 7(8)-en;2α-H	$C_{15}H_{20}O_4$	oil	+11.2	*L. vellereus*	*(58, 64)*
7.10	^{18}O;5-OH;7(8)-en; 2α,9α-H, ^{18}O-Isovellerol	$C_{15}H_{22}O^{18}O$	–	–	–	*(7)*
7.11	5-OH;7(8)-en;2α,9α-H, Isovellerol	$C_{15}H_{22}O_2$	–	+7.4	*L. vellereus, L. quietus*	*(7, 71)*
7.12	7α,8α-epoxy;5-OH; 2α,9α-H, Velutinal	$C_{15}H_{22}O_3$	oil	+39.6	*L. vellereus*	*(7, 37, 72)*
7.13	7α,8α-diOH;5-oxo; 2α,9α-H	$C_{15}H_{22}O_4$	238–240	+1.1	*L. vellereus*	*(43)*
7.14	7α,8β-diOH;5-oxo;2α, 9α-H	$C_{15}H_{22}O_4$	–	–	*L. vellereus,*	*(64)*
7.15	7α,8α,15-tri-OH; 5-oxo;2α,9α-H	$C_{15}H_{22}O_5$	oil	+1.5	*L. vellereus*	*(67)*
7.16	5-OH;7(8)-en;2α, 9α-H;12-acid Me ester, Rubrocinctal B	$C_{16}H_{22}O_4$	–	–	*L. rubrocinctus*	*(69)*
7.17	7α,8α-epoxy;5-OMe; 2α,9α-H, Methylvelutinal	$C_{16}H_{24}O_3$	–	–	–	*(36, 37)*
7.18	8α-OH;5α-OMe; 7(13)-en;2α,9α-H	$C_{16}H_{24}O_3$	oil	+236	–	*(17)*
7.19	10α-OAc;5-oxo; 7(8)-en;2α,9α-H	$C_{17}H_{22}O_4$	oil	+29.5	–	*(66)*
7.20	14-OAc;5-oxo; 7(8)-en;2α,9α-H	$C_{17}H_{22}O_4$	oil	+64.4	–	*(66)*
7.21	7α-OH,8α-OAc; 5-oxo;2α,9α-H	$C_{17}H_{24}O_5$	112–116	+37.5	–	*(43)*
7.22	7α-OH,8β-OAc; 5-oxo;2α,9α-H	$C_{17}H_{24}O_5$	oil	+71	–	*(64)*
7.23	10α-OCONHCOCCl$_3$; 5-oxo; 7(8)-en;2α,9α-H	$C_{18}H_{20}O_3NO_5$	203–208	+10.5	–	*(73)*

(continued on p. 92)

(continued from p. 91)

No.	Substituents, name	Formula	m.p. °C	$[\alpha]_D{}^{20}$	Source	Ref.
7.24	9α,10α-diOAc; 5-oxo;7(8)-en;2α-H	$C_{19}H_{24}O_6$	oil	–	–	(*58*)
7.25	7α-OH;8α,15-di-OAc; 5-oxo;2α,9α-H	$C_{19}H_{26}O_7$	oil	−64.3		(*67*)
7.26	Isovellerol dimer	?	–	–	–	(*7*)
7.27	5-(6′-oxostearoyloxy); 12-oxo;7(8)-en;2α,9α-H, 6′-Ketostearoyl-rubrocinctal A	$C_{33}H_{52}O_5$	–	–	*L. rubrocinctus*	(*69*)
7.28	7α.8α-epoxy;5α-(6′-oxostearoyloxy);2α, 9α-H, 6′-Ketostearoylvelutinal	$C_{33}H_{54}O_5$	oil	+54.8	*L. necator*, *L. rufus*, *L. trivialis*, *L. chrysorrheus* *L. mitissimus* *L. circellatus*,	(*9, 37, 38, 47, 69*)
7.29	8α-OH;5α-stearoyloxy;7(13)-en; 2α,9α-H	$C_{33}H_{56}O_4$	–	–	–	(*17*)
7.30	7α,8α-epoxy;5α-stearoyloxy;2α,9αH, Stearoylvelutinal	$C_{33}H_{56}O_4$	29	+55	*L. velutinus*, *L. vellereus*, *L. necator*, *L. piperatus*, *L. bertillonii*, *L. scrobiculatus*, *L. rufus* *L. trivialis*, *L. blennius*, *L. glaucescens*, *L. controversus*, *L. torminosus*, *L. circellatus*, *L. quietus*	(*8, 9, 36, 37, 38, 42, 47, 49, 60*)
7.31	5-(6′-oxostearoyloxy); 7(8)-en;2α,9α-H;12-acid Me ester, 6′-Ketostearoyl-rubrocinctal B	$C_{34}H_{54}O_6$	–	–	*L. rubrocinctus*	(*69*)

Part 8. Glutinopallane Sesquiterpenes

The only two known natural glutinopallane sesquiterpenes **8.2** and **8.3** were isolated from frozen fruiting bodies of *L. glutinopallens* extracted with CH_2Cl_2 (*74*). The two esters were separated by preparative HPLC on a RP-18 column with MeOH-H_2O, 95:5, the fatty acids being identified by GC after methanolysis, as palmitic and stearic acids. In the EIMS spectra of glutinopallal esters, the molecular ions at *m/z* 530 and 558, respectively, were very weak; by contrast, the MW 292 for the terpenoid moiety was corroborated by the presence of the base peak $(M\text{-}RCOO)^+$ at *m/z* 275 (in each spectrum). The cyclopropane and the 7,8-epoxy rings, both *cis* to the protons H-2 and H-9, relate the structures **8.2** and **8.3** to velutinal esters (**7.28**, **7.30**), while the carbomethoxy group at C-3 associate them to rubrocinctals **7.16** and **7.31**. The ^1H NMR features, which allowed distinction of stearoylglutinopallal (**8.3**) from structurally related stearoylvelutinal (**7.30**), were the two broad singlets at relatively low field (δ 1.73 and 1.64) which were attributed to the methyls attached to a sp^2 carbon, and the singlet at 3.70 ppm assigned to the carbomethoxy group at C-3. In addition, H-9 exhibited no coupling with H_2-10 due to contraction of the cyclopentane ring. In analogy with velutinal and rubrocintal esters, the ^1H NMR spectra of glutinopallal esters and O-methyl acetal **8.1** showed only one signal for the hemiacetal proton H-5 which NOE experiments located *trans* to the cyclopropane ring.

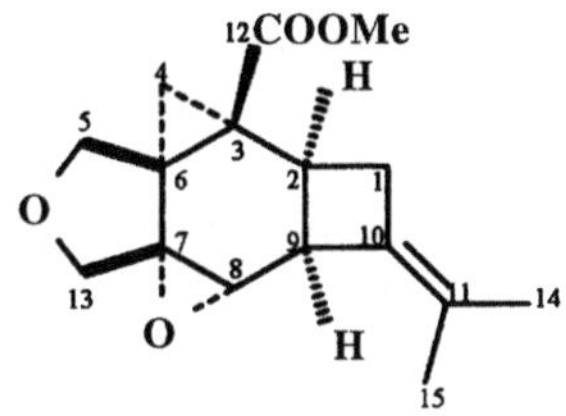

No.	Substituents, name	Formula	m.p. °C	$[\alpha]_D^{20}$	Source	Ref.
8.1	5α-OMe, Methylglutinopallal	$C_{17}H_{22}O_5$	–	–	–	*(74)*
8.2	5α-palmitoyloxy, Palmitoylglutinopallal	$C_{32}H_{50}O_6$	–	–	*L. glutinopallens*	*(74)*
8.3	5α-stearoyloxy, Stearoylglutinopallal	$C_{34}H_{54}O_6$	–	–	*L. glutinopallens*	*(74)*

Part 9. Isolactarane Sesquiterpenes

The fascinating lactone isolactarorufin (**9.4**), at first discovered in an EtOH extract of *L. rufus* (*75, 76, 77*) and later also in *L. vellereus* (*58*) and *L. necator*, is the only known isolactarane sesquiterpene so far isolated from *Lactarius* species. Other compounds listed in Part 9 are synthetic derivatives that were prepared for structure elucidation of the parent compound or for investigation of their biological properties.

The entire structure of isolactarorufin (**9.4**) was elucidated by spectroscopic methods (*76, 78*) and definitely proved by X-ray analysis of its *p*-bromobenzoate **9.8** (*77*). Important structural features of isolactarorufin are the *cis* configuration of the cyclopropane ring and the *trans* configuration of the C-3 hydroxy group with respect to the hydrogens at the ring junction. Notably, the configuration at C-3 is thus opposite to that of the related 3,8-dihydroxylactaranolide lactarorufin A (**11.30**) also isolated from the same extract of *L. rufus* (*75*) and may suggest a possible biosynthetic pathway. Indeed, the biogenesis of the isolactaranes is not well understood, but in view of the isolation of large amounts of velutinal esters **7.28** and **7.30** from hexane extracts of *L. rufus* (*8*), (*38*), it seems likely that both the isolactarane and the lactarane sesquiterpenes of *Lactarius* derive from the same marasmane precursor, albeit through different routes.

The ^{1}H NMR spectrum of **9.4** exhibited three methyl group singlets, one of them (C-12) resonating at lower field (δ 1.90 in Py-d_5) because of being geminal to the tertiary hydroxy group. The cyclopropane protons at C-5, as in the case of the marasmane sesquiterpenes, produced a pair of doublets (geminal coupling $J=5.5$ Hz) at δ 0.75 and 1.34. The methylene protons at C-13 gave an ABq system with $J=10$ Hz characteristic of saturated γ-lactone rings of this type. The relatively large coupling constant, $J=9.0$ Hz, of H-8 demonstrated the *trans* relationship with H-9; in fact, in the ^{1}H NMR spectrum of the synthetic stereoisomer 8-*epi*-isolactarorufin (**9.5**), the signal of H-8 is a broad singlet.

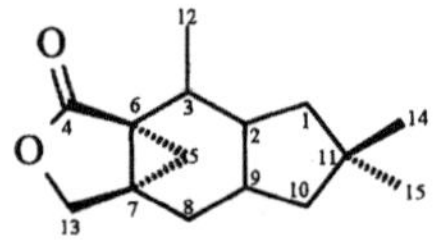

9.1 - 9.8

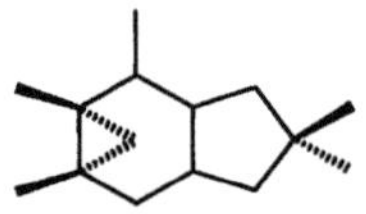

9.9 - 9.12

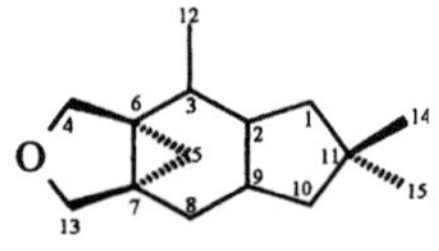

9.13 - 9.14

No.	Substituents, name	Formula	m.p. °C	$[\alpha]_D^{20}$	Source	Ref.
9.1	8-oxo;2(9)-en	$C_{15}H_{18}O_3$	178–179	+60	–	(*76, 78*)
9.2	8-OH;2(3)-en;9α-H	$C_{15}H_{20}O_3$	115–118	+80.16	–	(*76, 78*)
9.3	3β-OH;8-oxo;2α,9α-H	$C_{15}H_{20}O_4$	215–218	−19.6	–	(*76, 78*)
9.4	3β,8α-diOH;2α,9α-H, Isolactarorufin (Lactarorufin C)	$C_{15}H_{22}O_4$	191	+8.4	*L. rufus*, *L. vellereus*, *L. necator*	(*48, 58, 75, 76, 77, 78, 79*)
9.5	3β,8β-diOH;2α,9α-H	$C_{15}H_{22}O_4$	245–248	+4.3	–	(*76, 78*)
9.6	8α-OAc;2(3)-en;9a-H	$C_{17}H_{22}O_4$	91	−2.3	–	(*76, 78*)
9.7	3β-OH;8α-OAc;2α,9α-H	$C_{17}H_{24}O_5$	171–173	−18.2	–	(*76, 78*)

(continued on p. 95)

(continued from p. 94)

No.	Substituents, name	Formula	m.p. °C	$[\alpha]_D{}^{20}$	Source	Ref.
9.8	3β-OH;8α-OBz-*p*-Br; 2α,9α-H	$C_{22}H_{25}BrO_5$	208–210	–	–	*(77)*
9.9	4,8α,13-triOH;2(3)-en; 9α-H	$C_{15}H_{24}O_3$	180–184	+45.8	–	*(78)*
9.10	3β,4,8α,13-tetraOH; 2α,9α-H	$C_{15}H_{26}O_4$	154–7	−17.0	–	*(80)*
9.11	4,8α,13-triOAc;3(4)-en; 9α-H	$C_{21}H_{30}O_6$	oil	+15.9	–	*(78)*
9.12	3β-OH;4,8α,13-triOAc; 2α,9α-H	$C_{21}H_{32}O_7$	oil	−156.0	–	*(80)*
9.13	3β,4β,8α-triOH;2α, 9α-H	$C_{15}H_{24}O_4$	194–9	−42.0	–	*(80)*
9.14	3β-OH;4β.8α-diOAc; 2α,9α-H	$C_{19}H_{28}O_6$	oil	−347.0	–	*(80)*

Part 10. Lactarane Sesquiterpenes

Velleral (**10.6**), the first discovered lactarane dialdehyde, was isolated together with isovelleral (**6.2**) from *L. vellereus* by P.H. List (*53*) in 1969; however, only its partial structure was reported at that time. Later, during their pioneering work on the chemical constituents of *Lactarius* species Swedish authors reported full spectroscopic data and the complete structure of the same compound (**10.6**) (*54*, *55*) which, however, was erroneous regarding the stereochemistry at C-3. Eventually, revised structures of velleral (**10.6**), and the related lactones vellerolactone (**11.6**) and pyrovellerolactone (**11.7**) appeared in the literature in 1978 (*81*). Bicyclic lactarane sesquiterpenes may possess one or two double bonds within the cycloheptane ring; more often, however, they exhibit one double bond at 4(6) or contain a 4(6),7(8)- or 2(9),7(8)-diene system. Carbons 5 and 13 of natural lactaranes may carry either two carbonyl groups, (e.g. **10.6**) or one aldehyde and one hydroxy group (e.g. **10.10**), or two hydroxy groups (e.g. **10.14**). The most interesting compounds of this group are the dialdehydes chrysorrhedial (**10.2**), velleral (**10.6**), piperdial (**10.11**), and *epi*-piperdial (**10.12**) which, as well as lactardial (**14.2**) and isovelleral (**6.2**), impart a pungent taste to the human tongue and in general show antimicrobial, cytotoxic, antifeedant and mutagenic activities. As already discussed, they are rapidly (within seconds) formed from velutinal ester precursors in damaged fruiting bodies of *Lactarius*; therefore, they are considered to form a chemical defence system that is immediately activated after injury, to protect the mushrooms against parasites and predators. Moreover, these aldehydes are possibly the main compounds responsible for the intoxication arising from ingestion of the flesh of acrid *Lactarius* species. Depending on the species, a single dialdehyde or a mixture of two or more is present in fruiting bodies.

The ^{1}H NMR spectra of bicyclic lactaranes exhibit a typical pattern of two singlets and one doublet (*J ca.* 7 Hz), usually at δ 1.0–1.15 ppm, which are assigned to the geminal methyl groups at C-11, and to the C-12 methyl group, respectively. Indeed, the two singlets for the C-14 and C-15 methyl groups are characteristic of all the C-11 dimethyl substituted cyclopentane sesquiterpenes of *Lactarius*. The C-12 methyl group may be either *cis* or *trans* to H-2 and H-9 (when applicable) and this configuration cannot be established unambiguously on the basis of the coupling constants of H-3 with H-2 and H-4. Even the results of NOE experiments may be misleading (*55*), so that the configuration of velleral (**10.6**) and related compounds could be proved unambiguously only by total synthesis (*81*). On the other hand, NOE experiments along with molecular modelling (MM2) showed that H-6 and H_3-12 are *cis* on the cycloheptadiene ring of aldehydes **10.2** and **10.10** (*9*). The chemical shifts of aldehyde protons are indicative of whether the carbonyl group is conjugated or not to a double bond (conj. δ 9.25–9.50; nonconj. δ 9.70–9.90). On the other hand, olefinic protons in positions β to an aldehyde carbonyl group resonate at δ 6.45–7.10, significantly shifted downfield when compared with the non-conjugated protons (δ 5.76–6.10). The extended conjugated dienal system of compounds **10.2** and **10.10** shows a UV (CH_2Cl_2) absorption band at *ca* 315 nm (*9*) which is to be compared with the UV (EtOH) absorption at 245 nm for velleral (**10.6**) (*55*) and at 232 nm for piperdial (**10.11**) (*60*).

The absolute configuration of aldehydes **10.2** and **10.10** was established by hydride reduction to diol **10.13** which showed a positive CD curve and thus a positive skewness of the 2(9),7(8)-diene.

The lactarane sesquiterpenes **10.2**, **10.10**, **10.13**, containing the 2(9),7(8)-cycloheptadiene ring were submitted to conformational analysis by molecular mechanics and ^{1}H NMR spectroscopy (*9*). It was observed that the conformational mobility of each compound is practically restricted to the interconversion of the envelope forms of the cyclopentene ring; by contrast, essentially only a single conformation of the seven membered ring, with the 3-methyl group pseudo-equatorially oriented, is populated. This is due to the planarity of either the diene in **10.13** or the diene-carbonyl double bonds in **10.2**, **10.10**.

When the few γ-hydroxy aldehydes possessing the marasmane, isolactarane, or lactarane skeletons are compared, some interesting observations can be made. In principle, open γ-hydroxy aldehydes should exist as an equilibrium mixture with their two cyclic stereoisomeric hemiacetals. In fact, NMR spectra indicate that in common organic solvents isovellerol **7.11** exists as a mixture of approximately equal amounts of the three forms shown in Scheme 3 (*7*).

According to the above, the hemiacetal **9.13** obtained by hydride reduction of isolactarorufin **9.4** (*80*) should also exist in three forms. However, contrary to this expectation, the ^{1}H NMR spectrum of compound **9.13** ($CHCl_3$) revealed that only the 4β-OH form was present (Scheme 4).

This structure is stabilized by a strong intramolecular hydrogen bond between 3β and 4β OH groups as revealed by the X-ray structure (*80*). On the other hand, the ^{1}HNMR spectra of hydroxyaldehydes chrysorrheal (**10.10**), piperalol (**10.16**), and *epi*-piperalol (**10.17**) excluded the existence of hemiacetal forms. In fact, molecular modelling of **10.10** (*9*) clearly suggested that ring closure to the hemiacetal form would require a high conformational strain with loss of the resonance energy of the conjugated formyl group. It seems that the same factors are involved in the case of **10.16** and **10.17**.

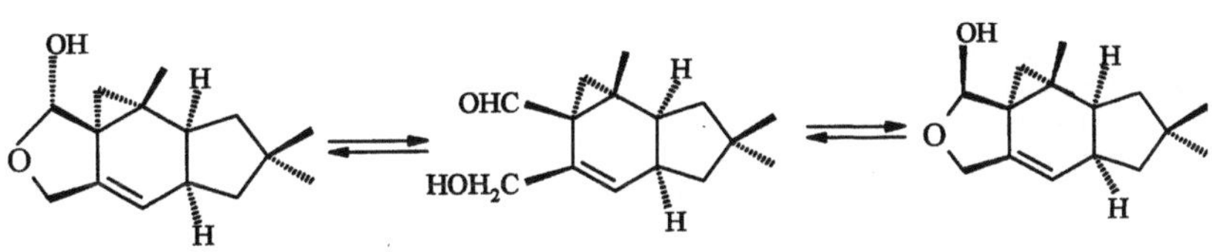

7.11

Scheme 3

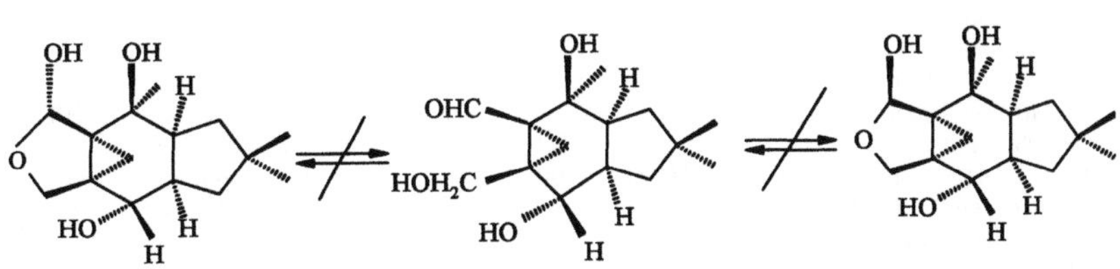

9,13

Scheme 4

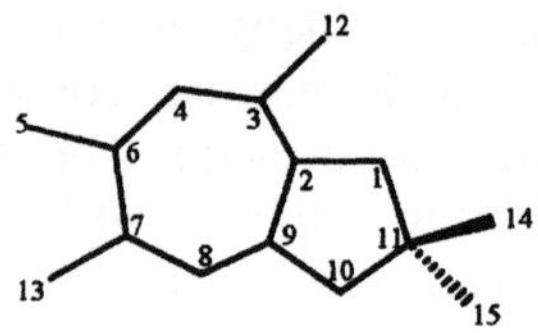

No.	Substituents, name	Formula	m.p. °C	$[\alpha]_D{}^{20}$	Source	Ref.
10.1	^{18}O;5,13-dioxo;4(6), 7(8)-dien;2α,3β,9α-H; ^{18}O-Velleral	$C_{15}H_{20}O$ ^{18}O	–	–	–	*(7)*
10.2	5,13-dioxo;2(9), 7(8)-dien;3α,6β-H; Chrysorrhedial	$C_{15}H_{20}O_2$	oil	+60.2	*L. chrysorrheus, L. scrobiculatus*	*(9)*
10.3	5,13-dioxo;3(12),7(8)-dien;2α,6α,9α-H	$C_{15}H_{20}O_2$	oil	+136.9	–	*(59)*
10.4	5,13-dioxo;3(12),7(8)-dien;2α,6β,9α-H	$C_{15}H_{20}O_2$	oil	+26.2	–	*(59)*
10.5	5,13-dioxo;4(6), 7(8)-dien;2α,3α,9α-H	$C_{15}H_{20}O_2$	–	–	–	*(81)*
10.6	5,13-dioxo;4(6),7(8)-dien;2α,3β,9α-H; Velleral	$C_{15}H_{20}O_2$	86.5–87.5	–25	*L. vellereus, L. piperatus, L. bertillonii, L. subvellereus, L. torminosus, L. circellatus, L. necator, L. rufus, L. pergamenus*	*(7, 17, 37, 40, 42, 47, 53, 55, 60, 81)*
10.7	9-OH;5,13-dioxo;4(6), 7(8)-dien; 2α,3β-H	$C_{15}H_{20}O_3$	131–133	–31.0	*L. vellereus*	*(7)*
10.8	^{18}O;13-OH;5-oxo;4(6), 7(8)-dien; 2α,3β,9α-H; ^{18}O-Vellerol	$C_{15}H_{22}O$ ^{18}O	oil	–	–	*(7)*
10.9	13-OH;5-oxo;4(6), 7(8)-dien;2α,3β,9α-H; Vellerol	$C_{15}H_{22}O_2$	oil	+149	*L. vellereus, L. bertillonii, L. piperatus, L. necator, L. torminosus, L. circellatus*	*(7, 42, 47, 60)*
10.10	5-OH;13-oxo;2(9), 7(8)-dien;3α,6β-H; Chrysorrheal (Scrobicalol)	$C_{15}H_{22}O_2$	oil	+29.6	*L. chrysorrheus, L. scrobiculatus*	*(9, 49)*
10.11	8α-OH;5,13-dioxo; 4(6)-en;2α,3β,7α,9α-H; Piperdial	$C_{15}H_{22}O_3$	oil	+77.0	*L. piperatus, L. torminosus Russula queletii*	*(17, 60)*
10.12	8α-OH;5,13-dioxo; 4(6)-en; 2α,3β,7β,9α-H; 7-*epi*-Piperdial	$C_{15}H_{22}O_3$	oil	–	*L. necator, L. circellatus*	*(17, 47)*
10.13	5,13-diOH;2(9),7(8)-dien; 3α,6β-H; Chrysorrhediol	$C_{15}H_{24}O_2$	53–54	+59.5	–	*(9, 49)*

(continued on p. 99)

(continued from p. 98)

No.	Substituents, name	Formula	m.p. °C	$[\alpha]_D^{20}$	Source	Ref.
10.14	5,13-diOH;4(6),7(8)-dien; 2α,3β,9α-H; Vellerdiol	$C_{15}H_{24}O_2$	oil	+203.0	*L. vellereus*	*(7, 82)*
10.15	3α,8α-epoxy;5,13-diOH; 6(7)-en; 2α,9α-H	$C_{15}H_{24}O_3$	96	+152		*(83)*
10.16	8α,13-diOH;5-oxo; 4(6)-en; 2α,3β,7α, 9α-H. Piperalol	$C_{15}H_{24}O_3$	oil	+57	*L. piperatus, L. torminosus Russula queletii*	*(60)*
10.17	8α,13-diOH;5-oxo; 4(6)-en; 2α,3β,7β, 9α-H; *epi*-Piperalol	$C_{15}H_{24}O_3$	oil	−62	*L. necator, L. circellatus*	*(47)*
10.18	5,8α,13-triOH; 2(3)-en;9α-H	$C_{15}H_{26}O_3$	oil	–	–	*(84)*
10.19	5,8α,13-triOH,4(6)-en; 2α,3β,7α,9α-H; Pipertriol	$C_{15}H_{26}O_3$	158–160	+62	–	*(60)*
10.20	5,8α,13-triOH;4(6)-en; 2α,3β,7β,9α-H;7-*epi*-Pipertriol	$C_{15}H_{26}O_3$	108–114	−39.0	*L. necator*	*(48)*
10.21	3α,5,8α,13-tetra-OH; 6(7)-en; 2α,9α-H	$C_{15}H_{26}O_4$	–	–	–	*(84)*
10.22	5,8α,13-triOH;2α, 3β,6α,7β,9α-H	$C_{15}H_{28}O_3$	–	–	–	*(48)*
10.23	5,8α,13-triOH;2α, 3β,6β,7β,9α-H	$C_{15}H_{28}O_3$	–	–	–	*(48)*
10.24	3α,5,8α,13-tetra-OH; 2α,9α-H	$C_{15}H_{28}O_4$	–	–	–	*(84)*
10.25	3α,8α-epoxy; 5,13-diOAc;6(7)-en; 2α,9α-H	$C_{19}H_{28}O_5$	oil	+85.6	–	*(83)*
10.26	5,8α,13-triOAc; 2(3)-en;9α-H	$C_{21}H_{32}O_6$	–	–	–	*(84)*
10.27	5,8α,13-triOAc;4(6)-en; 2α,3β,7β,9α-H	$C_{21}H_{32}O_6$	oil	−81.2	–	*(48)*
10.28	3α-OH;5,8α,13-tri-OAc; 6(7)-en;2α,9α-H	$C_{21}H_{32}O_7$	–	–	–	*(84)*
10.29	3α-OH;5,8α,13-tri-OAc; 2α,9αH	$C_{21}H_{34}O_7$	–	–	–	*(84)*
10.30	5,8α,13-triOCONHCOCCl$_3$; 4(6)-en;2α,3β,7β,9α-H	$C_{24}H_{26}Cl_9N_3O_6$	–	–	–	*(48)*
10.31	Vellerol dimer		–	–	*L. vellereus*	*(7)*

Part 11. 5-Lactaranolide Sesquiterpenes

5-Lactaranolides constitute the largest group of sesquiterpenoids isolated from *Lactarius* species. Thus part 11 includes more than 90 compounds, of which *ca.* 30 were isolated from extracts of *Lactarius*, the remaining ones being synthesized either for structure elucidation purposes or for testing their biological activity. Actually, in 1971 lactarorufin A was the first sesquiterpene isolated from *Lactarius* to have its hydroazulenic lactarane structure established (*83*). The position of the lactone carbonyl, initially assigned to C-13, was later revised (*85*). The true origin of *Lactarius* lactones of any class has long been debated. It is generally accepted now that lactones are formed in damaged fruiting bodies from velutinal ester precursors (e.g. **7.28** and **7.30**), through various steps where enzyme assistance seems indispensable, even if most details of the biosynthetic pathways are still unknown. Lactarane dialdehydes or hydroxyaldehydes (Part 10) have been proposed as the immediate precursors of the lactones. On the other hand, the possibility that some lactones are formed by oxidation of furans in air has been proved experimentally (*84*) and is further supported by the fact that for many lactones, the corresponding furan has been isolated (Parts 18 and 19) from the same mushrooms. Because of the common origin from velutinal esters, all (where applicable) lactaranolide and furanolactarane sesquiterpenes isolated from *Lactarius* possess a *cis* ring junction between the seven and five membered rings.

Most natural 5-lactaranolides possess one double bond in position 6(7) (lactarorufins **11.28**, **11.30**, **11.34** etc.), with exceptions where the unsaturation is at position 4(6) (**11.22**, **11.26**, **11.29**); a few conjugated (**11.7**, **11.12**, **11.16**) and deconjugated (**11.10**, **11.11**, **11.13**, **11.17**) dienes are also known. So far, only two examples of lactarane trienes have been reported (**11.1**, **11.3**).

Five lactaranolides carrying either 3α or 3β methyl group are known. Usually, the configuration of Me-3 is *cis* to the ring junction protons H-2 and H-9 when C-12 is geminal to a hydrogen and *trans* when it is geminal to a hydroxy group. It was suggested (*9*) that two alternative biosynthetic pathways may account for the two different configurations. Unambiguous assignment of the stereochemistry at C-3 to 5-lactaranolides required either single crystal X-ray analysis (*86*) or chemical evidence, while interpretation of the NMR data alone could be misleading.

Hydroxylated 5-lactaranolides may contain up to four OH groups, one of which is located on C-8 of most compounds. The orientation of this OH group is *cis* to H-9, as substantiated by the value of the coupling constant between H-8 and H-9 protons which usually ranges from 7 and 12 Hz and thus corresponds to a dihedral angle of ca. 180°. The magnitude of this dihedral angle, and accordingly the coupling constant, are different in lactarorufin A (**11.30**) and B (**11.41**). In these two compounds the *cis* OH groups at C-3 and C-8 form a strong intramolecular hydrogen bond that changes the overall molecular conformation, so that the signals of H-8 in the ^{1}H NMR spectra appear as broad singlets. (*9*, *86*). As expected, the ^{1}H NMR spectra of acetyl derivatives of lactarorufin A and B (**11.59** and **11.76**, respectively) exhibited $J_{8,9}=$ 9.8 and $J_{8,9}=10.6$, respectively.

In addition to the 8-OH group, other hydroxy groups may occur at carbons 2,3,4,7,13,14, and 15 of 5-lactaranolides. With exception of C-4 where the hydroxy group can be either α (**11.34**) or β (**11.36**), the OH groups on carbons C-2, C-3, and C-7 are *cis* to hydrogens H-2 and H-9. 13-Hydroxylactarolides (**11.16**, **11.18**, **11.40**, **11.42**), that are strongly suspected to be formed by air oxidation of the corresponding furans (Part 18), are fast equilibrating C-13 epimeric mixtures. This makes their separation impossible and causes doubling of many peaks in their ^{13}C-NMR spectra and of a few signals in their ^{1}H NMR spectra (*84*, *87*, *88*). Characteristic NMR features of the above γ-hydroxybutenolides are the ^{1}H (δ 5.85–6.10) and ^{13}C (δ_c 97.3–100) chemical shifts of the 13-methine, as well as the homoallylic coupling between H-13 and H-4 (*84*).

Three examples of 8-keto-5-lactaranolides (**11.1**, **11.2**, **11.4**) as well as three examples of epoxy-derivatives (**11.2**, **11.14**, **11.57**) were isolated.

Many 5-lactaranolides are nicely crystalline compounds; therefore, whenever possible, their structures were confirmed by X-ray analysis (*86*, *89*). Moreover, NMR spectra and molecular modelling of several representative lactones of this group (*86*, *90*) showed that the conformational mobility of each compound in solution is usually restricted to conversions of the five membered ring; by contrast, only a single conformation of the seven membered ring is practically populated at room temperature. This form can be described as a hinge conformation on which the C-3 methyl group is equatorially oriented (*86*, *90*). Thanks to the rigidity of the structures, orientation of the substituents on the central ring can therefore be inferred with confidence from the values of vicinal coupling constants. An exception is 3-*epi*-lactarorufin D (**11.33**), for which a conformation similar to lactarorufin A (**11.30**) was expected and, instead, dynamic NMR studies proved the existence of two conformations in $CDCl_3$ at rt. (*91*).

Simple circular dichroic method for the determination of absolute configuration of 5-substituted 2(5*H*)-furanones has been established (*172*). Using this method absolute configurations of lactarorufin A (**11.30**) and its 8-*epi*-derivative (**11.31**) were confirmed.

Thanks to the presence of hydroxy groups, simple reactions such as esterifications, oxidations, dehydration were extensively performed for structure elucidation of 5-lactaranolides; other relevant chemical transformations involved the conversion of the lactone into the furan ring and addition reactions to double bonds (hydrogenation, hydroboration, epoxidation, osmylation). A brief account of these reactions will be included in the chapter on chemical conversions of *Lactarius* sesquiterpenes.

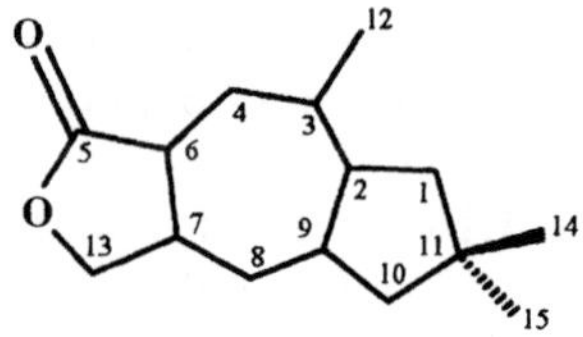

No.	Substituents, name	Formula	m.p. °C	$[\alpha]_D^{20}$	Source	Ref.
11.1	8-oxo;2(9),3(4),6(7)-trien, Lactarotropone	$C_{15}H_{16}O_3$	146–147	–	*L. pallidus*, *L. scrobiculatus*	(*59, 92*)
11.2	2,9-epoxy;8-oxo;3(4), 6(7)-dien	$C_{15}H_{16}O_4$	oil	–	*L. scrobiculatus*	(*92*)
11.3	2(3),6(7),8(9)-trien	$C_{15}H_{18}O_2$	103	–	*L. vellereus*	(*93, 94*)
11.4	3α-OH;8-oxo;2(9), 6(7)-dien	$C_{15}H_{18}O_4$	paste solid	–	*L. scrobiculatus*	(*92*)
11.5	15-D_1;3α,8α-epoxy; 6(7)-en;2α,9α-H	$C_{15}H_{19}DO_3$	–	–	–	(*95*)
11.6	4(6),7(8)-dien;2α,3β, 9α-H, Vellerolactone	$C_{15}H_{20}O_2$	oil	+364	*L. vellereus*, *L. pergamenus* *L. bertillonii*	(*7, 42, 81, 96*)
11.7	3(4),6(7)-dien;2α, 9α-H, Pyrovellerolactone	$C_{15}H_{20}O_2$	41–44	–73	*L. vellereus* *L. pergamenus*	(*7, 81, 94, 96*)
11.8	3α-OH;6(7),8(9)-dien; 2α-H	$C_{15}H_{20}O_3$	75–78	+334.5	–	(*97*)
11.9	8α-OH;1(2),6(7)-dien; 3α,9β-H	$C_{15}H_{20}O_3$	oil	+14.5	–	(*98*)
11.10	8α-OH;1(2),6(7)-dien; 3α,9β-H	$C_{15}H_{20}O_3$	–	–	–	(*98*)
11.11	8α-OH;2(3),6(7)-dien; 9α-H	$C_{15}H_{20}O_3$	85–90	–3.8	*L. necator*, *L. pallidus*, *L. controversus*, *L. pergamenus*, *L. vellereus*, *L. scrobiculatus*, *L. torminosus*, *L. blennius*, *L. turpis*, *L. vietus*, *L. glyciosmus*, *L. helvus*, *L. spinosulus*, *L. quietus*, *L. subdulcis*, *L. thejogalus*	(*44, 56, 85, 92, 99, 100, 101, 102*)

(continued on p. 103)

(continued from p. 102)

No.	Substituents, name	Formula	m.p. °C	$[\alpha]_D^{20}$	Source	Ref.
11.12	8α-OH;3(4),6(7)-dien; 2α,9α-H	$C_{15}H_{20}O_3$	–	–	*L. controversus, L. pergamenus, L. vellereus, L. torminosus, L. blennius, L. necator, L. vietus, L. glyciosmus, L. spinosulus, L. quietus*	(*44, 85, 100*)
11.13	8α-OH;3(12),6(7)-dien; 2α,9α-H	$C_{15}H_{20}O_3$	114–115	+26.9	*L. necator, L. controversus, L. pergamenus, L. vellereus, L. scrobiculatus, L. torminosus, L. blennius, L. turpis, L. vietus, L. helvus, L. spinosulus, L. quietus, L. thejogalus*	(*44, 92, 100*)
11.14	3α,8α-epoxy; 6(7)-en;2α,9α-H	$C_{15}H_{20}O_3$	97	+132	*L. necator*	(*83, 97*)
11.15	3α-OH;8-oxo; 6(7)-en;2α,9α-H	$C_{15}H_{20}O_4$	110	−37.6	–	(*83*)
11.16	4β,13-diOH;6(7),8(9)-dien;2α,3β-H Subvellerolactone	$C_{15}H_{20}O_4$	oil	–	*L. subvellereus*	(*88*)
11.17	8α,13-diOH;2(3),6(7)-dien;9α-H, Blennin B	$C_{15}H_{20}O_4$	oil	+70.78	*L. blennius*	(*103*)
11.18	3α,13-diOH;8-oxo; 6(7)-en;2α,9α-H	$C_{15}H_{20}O_5$	176–178	+27.86	–	(*84*)
11.19	3α,8α-diO(SO); 6(7)-en;2α,9α-H	$C_{15}H_{20}O_5S$	133	–	–	(*83*)
11.20	8(9)-en;2α,3α,6α.7α-H	$C_{15}H_{22}O_2$	oil	−9.7	–	(*104*)
11.21	8β-OH;6(7)-en;9α-H	$C_{15}H_{22}O_3$	–	–	–	(*56*)
11.22	8α-OH;4(6)-en;2α,3β,7β, 9α-H, Lactarorufin N	$C_{15}H_{22}O_3$	oil	−15.1	*L. necator, L. torminosus, L. circellatus, L. vietus, L. glyciosmus, L. thejogalus*	(*44, 47, 85, 100, 104*)
11.23	8α-OH;6(7)-en;2β, 3α,9α-H	$C_{15}H_{22}O_3$	oil	−12	–	(*98*)
11.24	8α-OH;6(7)-en;2β, 3β,9α-H	$C_{15}H_{22}O_3$	166–168	+10.8	–	(*98*)
11.25	8-oxo;2α,3α,6α,7α,9α-H	$C_{15}H_{22}O_3$	119–121	−36.24	–	(*85, 104*)
11.26	8α-OH;4(6)-en;2α, 3β,7α,9α-H, Blennin A	$C_{15}H_{22}O_3$	oil	+49.9	*L. blennius, L. torminosus, L. glyciosmus, L. thejogalus*	(*44, 45, 87, 100, 101, 102, 103, 105*)

(continued on p. 104)

(continued from p. 103)

No.	Substituents, name	Formula	m.p. °C	$[\alpha]_D^{20}$	Source	Ref.
11.27	8α-OH;6(7)-en;2α, 3α,9α-H	$C_{15}H_{22}O_3$	116–122	+70.0	*L. necator*	(*104, 106*)
11.28	8α-OH;6(7)-en;2α, 3β,9α-H	$C_{15}H_{22}O_3$	154–160	+61.0	*L. vellereus, L. blennius, L. turpis, L. vietus, L. glyciosmus, L. spinosulus, L. subdulcis*	(*44, 100, 104, 106*)
11.29	2α,8α-diOH;4(6);en; 3β,7α,9α-H, Blennin D	$C_{15}H_{22}O_4$	–	+51	*L. blennius*	(*90, 105*)
11.30	3α,8α-diOH;6(7)-en; 2α,9α-H, Lactarorufin A	$C_{15}H_{22}O_4$	156–158	+7	*L. rufus, L. necator, L. torminosus, L. vellereus, L. blennius, L. pallidus, L. trivialis, L. spinosulus, L. quietus, L. mitissimus, L. controversus, L. subdulcis*	(*48, 56, 58, 64, 75, 79, 83, 86, 89, 101, 102, 107, 108, 109, 110*)
11.31	3α,8β-diOH;6(7)-en; 2α,9α-H;8-*epi*-Lactarorufin A	$C_{15}H_{22}O_4$	87	−7.8	–	(*83*)
11.32	3α,8β-diOH;6(7)-en;2α, 9β-H;8-*epi*-9-*epi*-Lactarorufin A	$C_{15}H_{22}O_4$	192	+2.5	–	(*111*)
11.33	4α,8α-diOH;6(7)-en;2α, 3α,9α-H 3-*epi*-Lactarorufin D	$C_{15}H_{22}O_4$	165–175	+6.3	–	(*91*)
11.34	4α,8α-diOH;6(7)-en;2α, 3β,9α-H, Lactarorufin D	$C_{15}H_{22}O_4$	160–162	+93	*L. necator, L. torminosus*	(*79, 86*)
11.35	4β,8α-diOH;6(7)-en; 2α,3α,9α-H	$C_{15}H_{22}O_4$	–	+39.8	–	(*91*)
11.36	4β,8α-diOH;6(7)-en;2α, 3β,9α-H, Lactarorufin E	$C_{15}H_{22}O_4$	125–130	+58	*L. necator, L. torminosus*	(*79, 86, 91*)
11.37	7α,8α-diOH;4(6)-en; 2α,3β,9α-H, Sardonialactone A	$C_{15}H_{22}O_4$	163.5–164.5	−47.8	*L. necator, L. pallidus, L. rufus*	(*56, 79, 86, 112*)
11.38	8α,14-diOH;4(6)-en; 2α,3β,7α,9α-H	$C_{15}H_{22}O_4$	144	–	*L. torminosus*	(*45, 101, 102, 113*)
11.39	3α-OH;8-oxo;2α, 6α,7α,9α-H	$C_{15}H_{22}O_4$	148	+18.56	–	(*83, 85*)
11.40	3α,8α,13-triOH;6(7)-en; 2α,9α-H, Lactarolide A	$C_{15}H_{22}O_5$	153–155	+59.84	*L. blennius, L. pallidus, L. scrobiculatus, L. mitissimus,*	(*84, 87, 109, 114*)
11.41	3α,8α,15-triOH;6(7)-en; 2α,9α-H, Lactarorufin B	$C_{15}H_{22}O_5$	213	+24	*L. rufus, L. mitissimus*	(*75, 87, 95, 107, 109, 110*)

(continued on p. 105)

(continued from p. 104)

No.	Substituents, name	Formula	m.p. °C	$[\alpha]_D{}^{20}$	Source	Ref.
11.42	3α,8α,13,15-tetraOH; 6(7)-en; 2α,9α-H	$C_{15}H_{22}O_6$	oil	+29.0	*L. mitissimus*	*(87, 109)*
11.43	8α-OH;2α,3α,6α,7α,9α-H	$C_{15}H_{24}O_3$	110–115	+62.7	–	*(85, 104)*
11.44	8β-OH;2α,3α,6α,7α, 9α-H	$C_{15}H_{24}O_3$	–	–	–	*(104)*
11.45	3α,8α-diOH;2α,6α,7α, 9α-H	$C_{15}H_{24}O_4$	99–101	+60.5	–	*(75)*
11.46	3α,8α-diOH;2α,6β,7β, 9α-H	$C_{15}H_{24}O_4$	136	+20.1	–	*(75)*
11.47	3α,8α-epoxy;15-OMs; 6(7)-en;2α,9α-H	$C_{16}H_{22}O_6S$	206	+89.8	–	*(95)*
11.48	8α-OMs;4(6)-en;2α, 3β,7α,9α-H	$C_{16}H_{24}O_5S$	–	–	–	*(105)*
11.49	8α-OAc;3(4),6(7)-dien; 2α,9α-H	$C_{17}H_{22}O_4$	115–120	–156.0	–	*(75, 85, 114)*
11.50	8α-OAc;3(12),6(7)-dien;2α,9α-H	$C_{17}H_{22}O_4$	oil	+11.2	–	*(44)*
11.51	3α,4α-epoxy;8α-OAc; 6(7)-en;2α,9α-H	$C_{17}H_{22}O_5$	141–142	–	–	*(115)*
11.52	8α-OAc;4(6)-en;2α, 3β,7β,9α-H	$C_{17}H_{24}O_4$	oil	–52	–	*(85)*
11.53	8α-OAc;4(6)-en;2α, 3β,7α,9α-H	$C_{17}H_{24}O_4$	oil	+63.8	–	*(103)*
11.54	8α-OAc;6(7)-en; 2α,3α,9α-H	$C_{17}H_{24}O_4$	oil	–122	–	*(104, 106)*
11.55	8α-OAc;6(7)-en; 2α,3β,9α-H	$C_{17}H_{24}O_4$	oil	–	–	*(106)*
11.56	3α-OEt;8-oxo;6(7)-en; 2α,9α-H	$C_{17}H_{24}O_4$	90	–33.6	–	*(116)*
11.57	3β,8β-epoxy;13α-Et; 13β-OH;6(7)-en;2α,9α-H, Subvellerolactone C	$C_{17}H_{24}O_4$	–	–	*L. subvellereus*	*(117)*
11.58	2α-OH;8α-OAc;4(6)-en; 3β;7α,9α-H	$C_{17}H_{24}O_5$	–	–	–	*(105)*
11.59	3α-OH;8α-OAc;6(7)-en; 2α,9α-H	$C_{17}H_{24}O_5$	114–115	–21.0	–	*(75, 87, 114)*
11.60	3α-OH;8β-OAc;6(7)-en; 2α,9α-H	$C_{17}H_{24}O_5$	178	–34.6	–	*(83)*
11.61	3α-OH;8β-OAc;6(7)-en; 2α,9β-H	$C_{17}H_{24}O_5$	195	+7.0	–	*(111)*
11.62	4α-OH;8α-OAc;6(7)-en; 2α,3β,9α-H	$C_{17}H_{24}O_5$	–	–	–	*(91)*
11.63	4β-OH;8α-OAc;6(7)-en; 2α,3α,9α-H	$C_{17}H_{24}O_5$	–	–	–	*(91)*
11.64	7α-OH;8α-OAc;4(6)-en; 2α,3β,9α-H	$C_{17}H_{24}O_5$	200–201	–	–	*(112)*
11.65	3α,13-diOH;8α-OAc; 6(7)-en;2α,9α-H	$C_{17}H_{24}O_6$	143–147	–	–	*(84, 87, 114)*
11.66	8α-OH;3α-OEt;6(7)-en; 2α,9α-H	$C_{17}H_{26}O_4$	oil	–17	–	*(84, 116)*
11.67	8β-OH;3α-OEt;6(7)-en; 2α,9β-H	$C_{17}H_{26}O_4$	112	+5.0	–	*(118)*

(continued on p. 106)

(continued from p. 105)

No.	Substituents, name	Formula	m.p. °C	$[\alpha]_D^{20}$	Source	Ref.
11.68	8α-OAc;2α,3α,6α,7α,9α-H	$C_{17}H_{26}O_4$	129–131	+39.7	–	*(85, 104)*
11.69	8α,13-diOH;3α-OEt; 6(7)-en;2α,9α-H, 3-O-Ethyl-lactarolide A	$C_{17}H_{26}O_5$	171–173	+18.5	*L. blennius*	*(84)*
11.70	3α-OH;8α-OAc;2α,6, 7,9α-H	$C_{17}H_{26}O_5$	132–136	+28.6	–	*(75)*
11.71	8α-OH;3α-OAc;2α,6,7, 9α-H	$C_{17}H_{26}O_5$	129	+28.5	–	*(75, 106)*
11.72	8α,15-diOAc;3(4), 6(7)-dien;2α,9α-H	$C_{19}H_{24}O_6$	oil	–	–	*(95)*
11.73	3α,8α-diOAc;6(7)-en; 2α,9α-H	$C_{19}H_{26}O_6$	oil	−12.3	–	*(75)*
11.74	3α-OH;8α,13α-diOAc; 6(7)-en;2α,9α-H	$C_{19}H_{26}O_7$	oil	−75.9	–	*(84, 109)*
11.75	3α-OH;8α,13β-diOAc; 6(7)-en;2α,9α-H	$C_{19}H_{26}O_7$	oil	+13.17	–	*(84, 109)*
11.76	3α-OH;8α,15-diOAc; 6(7)-en;2α,9α-H	$C_{19}H_{26}O_7$	oil	+5.7	–	*(95)*
11.77	3α,13-diOH;8α,15-diOAc; 6(7)-en;2α,9α-H	$C_{19}H_{26}O_8$	–	–	–	*(87)*
11.78	3α-OEt;8α-OAc; 6(7)-en;2α,9α-H	$C_{19}H_{28}O_5$	110	+3.3	–	*(116)*
11.79	3α-OEt;8β-OAc; 6(7)-en;2α,9β-H	$C_{19}H_{28}O_5$	oil	−3.8	–	*(118)*
11.80	3α,8α-diOAc;2α,6,7, 9α-H	$C_{19}H_{28}O_6$	oil	+21.3	–	*(75)*
11.81	3α,8α15-triOAc; 6(7)-en;2α,9α-H	$C_{21}H_{28}O_8$	oil	–	–	*(95)*
11.82	3α-OH;8α,13α,15-triOAc; 6(7)-en;2α,9α-H	$C_{21}H_{28}O_9$	oil	−25.9	–	*(87)*
11.83	3α-OH;8α,13β,15-triOAc; 6(7)-en;2α,9α-H	$C_{21}H_{28}O_9$	oil	+37.1	–	*(87)*
11.84	3α-OEt;8α,13-diOAc; 6(7)-en;2α,9α-H	$C_{21}H_{30}O_7$	–	–	–	*(84)*
11.85	3α,8α-epoxy;15-(OBz-*p*-Br);6(7)-en;2α, 9α-H	$C_{22}H_{23}BrO_5$	162–165	+80.5	–	*(95, 107)*
11.86	3α,8α-diOH; 15-(OBz-*p*-Br);6(7)-en;2α, 9α-H	$C_{22}H_{25}BrO_6$	89–92	+2.8	–	*(95)*
11.87	3α,8α-epoxy;15-OTs; 6(7)-en;2α,9α-H	$C_{22}H_{26}O_6S$	139	+81.6	–	*(95)*
11.88	3α,8α-diOH;15-OTs; 6(7)-en;2α,9α-H	$C_{22}H_{28}O_7S$	155	+9.1	–	*(95)*
11.89	3α-OH;8α,15-di(OBz-*p*-Br);6(7)-en; 2α,9α-H	$C_{29}H_{28}Br_2O_7$	114–116	−13.8	–	*(95)*
11.90	3α-OH;8α,15-diOTs; 6(7)-en;2α,9α-H	$C_{29}H_{34}O_9S_2$	–	–	–	*(115)*
11.91	3α,-OH;8α-stearoyloxy; 6(7)-en;2α,9α-H	$C_{33}H_{56}O_5$	oil	−17.7	–	*(115)*

Part 12. 5-Lactaranolide Derivatives

5-Lactaranolide derivatives included in Part 12 are produced by silica gel degradation of velutinal esters (**7.28**, **7.30**), methylvelutinal (**7.17**) or free velutinal itself (for a discussion of these transformations see the chapter on chemical interconversions of *Lactarius* sesquiterpenes). The compounds upon acidification underwent aromatization to form furanoid derivatives (see Part 18). Under these conditions dehydration reactions took place and also dienes were formed (*12*).

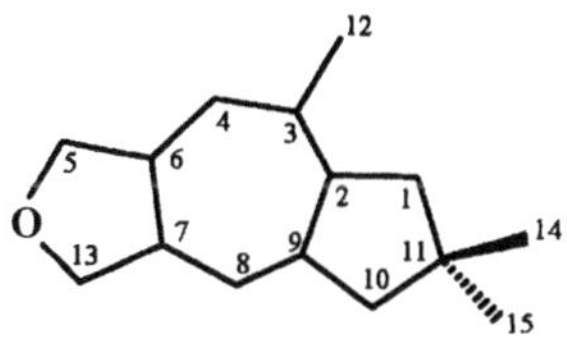

No.	Substituents, name	Formula	m.p. °C	$[\alpha]_D^{20}$	Source	Ref.
12.1	3α,8α-epoxy;5-OH; 6(7)-en;2α,9α-H	$C_{15}H_{22}O_3$	oil	+98.1	–	(*12*)
12.2	3α,5,8α-triOH;6(7)-en; 2α,9α-H	$C_{15}H_{24}O_4$	130–132	−1.6	–	(*12*)
12.3	5-OMe;4(6),7(8)-dien; 2α,3α,9α-H	$C_{16}H_{24}O_2$	–	–	–	(*81*)
12.4	3α,8α-epoxy;5β-OMe; 6(7)-en;2α,9α-H	$C_{16}H_{24}O_3$	54–58	+30.0	–	(*12*)
12.5	8α-OH;5β-OMe;2(3), 6(7)-dien;2α,9α-H	$C_{16}H_{24}O_3$	oil	+15.6	–	(*12*)
12.6	8α-OH;5β-OMe;3(12), 6(7)-dien;2α,9α-H	$C_{16}H_{24}O_3$	oil	+18.0	–	(*12*)
12.7	3α,8α-diOH;5-OMe; 6(7)-en;2α,9α-H	$C_{16}H_{26}O_4$	121–124	+15.3	–	(*12*)
12.8	5,8α-diOH;3α-OMe; 6(7)-en;2α,9α-H	$C_{16}H_{26}O_4$	oil	−2.4	–	(*12*)
12.9	8α-OH;3α,5α-diOMe; 6(7)-en;2α,9α-H	$C_{17}H_{28}O_4$	62–64	−51.0	–	(*12*)

Part 13. Rearranged 5-Lactaranolide Sesquiterpenes

Compounds included in Part 13 are synthetic derivatives, obtained from blennin A (**11.26**) (*105*) and lactarorufin A derivative **11.25** (*104*), respectively.

The ^{1}H NMR spectrum of compound **13.1** was similar to that of pyrovellerolactone (**11.7**) except for the signals of the C-13 methylene group. These two protons did not exhibit an isolated AB quartet but were further coupled with H-7, while the value of the geminal coupling constant (9 Hz) was much smaller than that typical of 6,7-en-5-lactaranolides. Moreover, decoupling experiments showed that the proton H-8 resonated at higher field (*ca.* δ 1) which suggested its location on a cyclopropane ring.

In the ^{1}H NMR spectrum of compound **13.2**, the C-13 methylene group doublet was shifted up field (δ 3.58, $J=6.5$ Hz) in comparison with the non-rearranged lactone **11.44**; in its acetyl derivative **13.3** the signal was shifted downfield (δ 4.03 ppm), as expected.

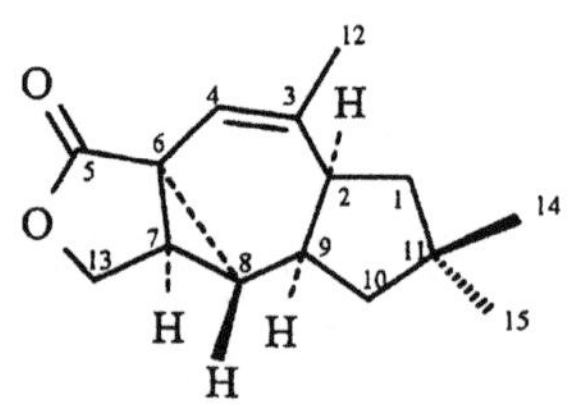

13.1

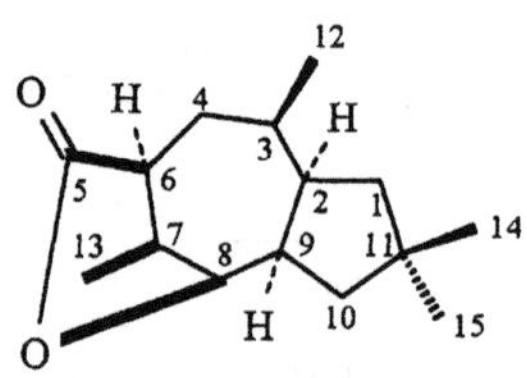

13.2-13.3

No.	Substituents, name	Formula	m.p. °C	$[\alpha]_D^{20}$	Source	Ref.
13.1	(see formula)	$C_{15}H_{20}O_2$	oil	−4.2	–	(*105*)
13.2	13-OH	$C_{15}H_{24}O_3$	–	–	–	(*104*)
13.3	13-OAc	$C_{17}H_{26}O_4$	oil	−39.4	–	(*104*)

Part 14. 8,9-Seco-5-lactaranolide Sesquiterpenes and Derivatives

Three sesquiterpenes (**14.1**, **14.2**, and **14.3**) of this group are considered genuine metabolites of *Lactarius* and blennin C (**14.3**) is apparently one of the most frequently isolated. The structure initially assigned to blennin C had the lactone carbonyl at C-13 (*50, 99*), but was later revised to **14.3** on the basis of chemical and spectroscopic evidence (*103*). Lactardial (**14.2**), formally a 1,4-dialdehyde possessing a pungent taste and antimicrobial activity, is formed from velutinal esters (**7.28** and **7.30**) both through enzymatic and chemical routes (*12*).

It is remarkable that in the ^{1}H NMR spectra of compounds **14.1**, **14.2**, **14.3**, **14.6**, and **14.7**, the methylene protons at C-1 and C-10 form a four proton broad singlet at *ca.* δ 2.10 and that the vinylic proton at C-9 is a broad singlet at *ca.* δ 5.25. Characteristic for lactones **14.1** and **14.3** is also the two-protons broad singlet at *ca* δ 4.90 assigned to the C-13 methylene.

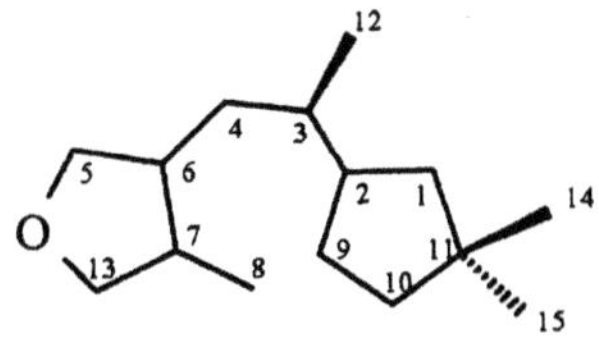

No.	Substituents, name	Formula	m.p. °C	$[\alpha]_D^{20}$	Source	Ref.
14.1	5,8-dioxo;2(9), 6(7)-dien	$C_{15}H_{20}O_3$	–	−8.97	*L. scrobiculatus*	(*49, 98*)
14.2	5-OH;8-oxo;2(9), 6(7)-dien, Lactardial	$C_{15}H_{22}O_3$	oil	+23.8	*L. piperatus, L. scrobiculatus, L. torminosus, L. chrysorrheus, L. circellatus, L. necator*	(*9, 12, 47, 49, 60*)
14.3	8-OH;5-oxo;2(9), 6(7)-dien, Blennin C (Lactaronecatorin A)	$C_{15}H_{22}O_3$	oil	−5.6	*L. necator, L. blennius, L. pallidus, L. vellereus, L. scrobiculatus, L. torminosus, L. circellatus, L. trivialis, L. turpis, L. vietus, L. glyciosmus, L. subdulcis, L. thejogalus, L. chrysorrheus*	(*9, 44, 47, 49, 50, 56, 99, 100, 101, 102, 103*)
14.4	5-oxo;6(7)-en	$C_{15}H_{24}O_2$	–	–	–	(*103*)
14.5	5-oxo	$C_{15}H_{26}O_2$	–	–	–	(*103*)
14.6	5α-OMe;8-oxo;2(9), 6(7)-dien	$C_{16}H_{24}O_3$	oil	+7.9	–	(*12*)
14.7	8-OAc;5-oxo;2(9), 6(7)-dien	$C_{17}H_{24}O_4$	oil	−13.3	–	(*99*)

Part 15. Norlactarane Sesquiterpenes

The structurally interesting compounds **15.1** and **15.2** are the only known examples of this class. The molecules have a plane of symmetry and their ^{1}H NMR spectra are therefore very simple, containing only six and seven signals, respectively. Both geminal methyl groups produce a six protons singlet.

8-Norlactarane sesquiterpene **15.1** is considered to be biosynthetically related to α,β-epoxyketone **11.2**, and it has been suggested that expulsion of the C-8 carbonyl group from the lactarane skeleton occurs *via* a benzylic-like rearrangement, followed by a decarboxylative aromatization (*92*) (Scheme 5).

Compound **15.2** is a minor product of preparative autoxidation of isovelleral (**6.2**) (*62*).

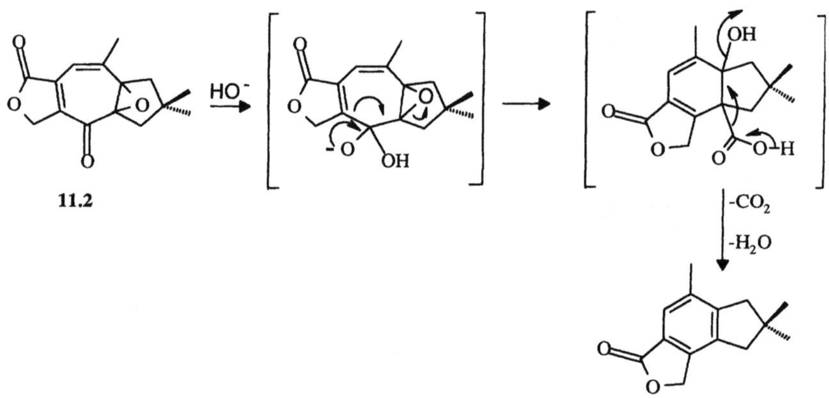

15.1

Scheme 5

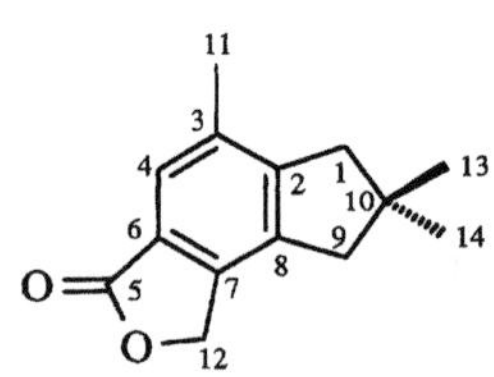

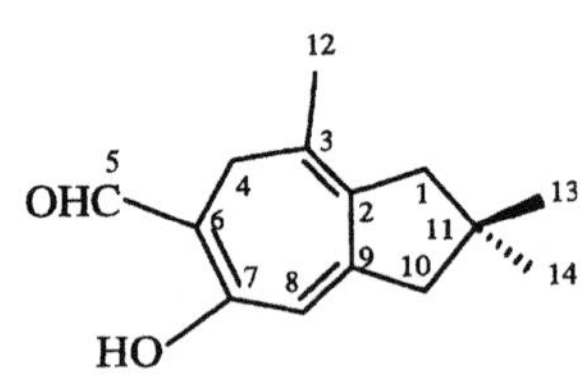

15.1 **15.2**

No.	Substituents, name	Formula	m.p. °C	$[\alpha]_D^{20}$	Source	Ref.
15.1	(See formula)	$C_{14}H_{16}O_2$	93–96	–	*L. scrobiculatus*	(*92*)
15.2	(See formula)	$C_{14}H_{18}O_2$	–	–	–	(*62*)

Part 16. 13-Lactaranolide Sesquiterpenes

13-Lactaranolides are much less numerous than the corresponding 5-lactaranolides and only seven members of this group (**16.1**, **16.2**, **16.6**, **16.7**, **16.8**, **16.14**, and **16.24**) have been isolated from *Lactarius* species. However, the latter three compounds are strongly suspected of being artifacts, epoxide **16.8** being produced by oxidation in air of lactaroscrobiculide A **16.2**, while **16.14**, and **16.24** may be formed by oxidation in air of furans **18.14** and **18.26**, respectively. The other lactones included in Part 16 are synthetic derivatives, most of them having been obtained by oxidation of the corresponding lactarane furans (Part 18). The extremely labile triene-enol lactone **16.1** was found to be involved in the rapid yellowing of the milky juice and flesh of *L. chrysorrheus* and *L. scrobiculatus*, as revealed by the strong UV absorption at 370.4 nm. Chrysorrhelactone (**16.1**), as well as the related aldehydes chrysorrhedial (**10.2**) and chrysorrheal (**10.10**), were particularly unstable as a neat liquid and had to be extracted from minced fruiting bodies with hexane in the dark. Moreover, due to the rapid degradation of the compounds, spectral data had to be immediately recorded after chromatographic separations (*9*). The ^{1}H NMR spectra of 6(7)-en-13-lactaranolides are very similar to those of isomeric 5-lactaranolides (Part 11); however, differentiation between the two groups can easily be achieved by NOE experiments involving the H_2-5 and H_2-13 methylene protons, respectively. For example, a strong NOE is observed between the H-8 and H_2-13 protons of 5-lactaranolides, whereas on irradiation of the H_2-5 protons in 13-lactaranolides, an enhancement is observed for the signals of the protons at C-4. For 7(8)-en-13-lactaranolides, e.g. lactaroscrobiculide A (**16.2**), the entire spin system from H_2-5 through H_3-12 could unambiguously be established by COSY experiments. Moreover, diagnostic for the Δ7(8) – unsaturation was the signal of H-8, ranging from δ 6.80 in compound **16.1** to δ 6.91 in **16.8** (*9*, *119*). Chrysorrhelactone **16.1** had an additional vinylic proton at δ 6.18 indicative of the double bond in the furanone ring. The relative configuration of lactones **16.2** and **16.8** was proved by NOE experiments and molecular modelling (MM2), while the absolute configuration was established by chemical correlation with diol **10.13**. Biosynthetic considerations suggested for lactones **16.1** and **16.2** the same absolute configuration as that of sesquiterpenes **10.2** and **10.10** (*9*).

Molecular mechanics calculations (MM2) (*9*), in addition to ^{1}H-NMR data, proved that for all the related sesquiterpenes **10.2**, **10.10**, **10.13**, **16.2**, and **16.8** a single conformation of the cycloheptadiene ring is practically the only one; however, the geometry of the global minima of lactaranes **10.2**, **10.10**, and **10.13** is different from those of 13-lactaranolides **16.2** and **16.8**. In fact, the C-3 Me group is pseudoequatorially oriented in the former three compounds, while it is pseudoaxial in the latter two, possibly due to additional constraints by the γ-lactone ring.

The simulated ^{13}C NMR spectrum of lactone **16.6** isolated from *L. vellereus* (*94*) suggested that the configuration at C-3 was opposite to that of isomeric lactaroscrobiculide A (**16.2**). This stereochemistry was established unequivocally by correlation of sesquiterpene **16.6** with 3-deoxy-3-*epi*-lactarorufin A (**11.28**), as shown in Scheme 6 (*94*).

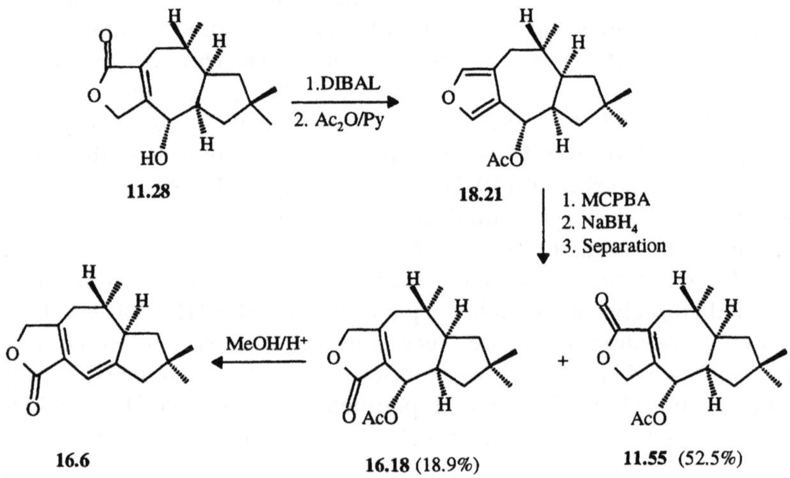

Scheme 6

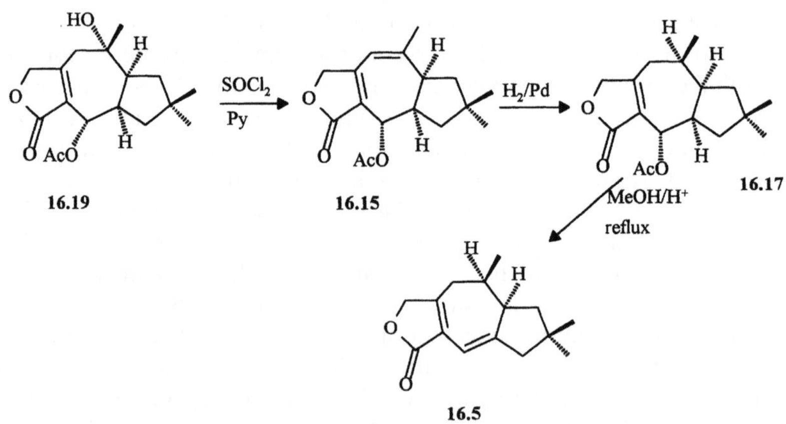

Scheme 7

For comparison, 3-deoxylactaroscrobiculide B (**16.5**), the C-3 epimer of the natural lactone **16.6**, was synthesized from lactone **16.17**, as shown in Scheme 7 (*94*).

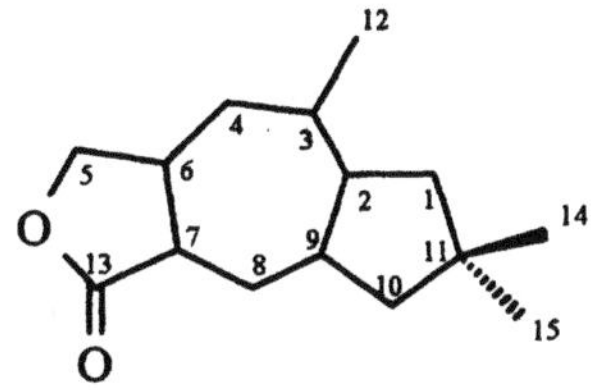

No.	Substituents, name	Formula	m.p. °C	$[\alpha]_D^{20}$	Source	Ref.
16.1	2(9),5(6),7(8)-trien; 3α-H, Chrysorrhe-lactone	$C_{15}H_{18}O_2$	oil	−30.4	*L. chrysorrheus*, *L. scrobiculatus*	(*9*)
16.2	2(9),7(8),dien; 3α,6β-H, Lactaro-scrobiculide A	$C_{15}H_{20}O_2$	–	–	*L. chrysorrheus*, *L. scrobiculatus*	(*9, 49, 50*)
16.3	3(4),6(7)-dien;2α, 9α-H	$C_{15}H_{20}O_2$	90–92	–	–	(*81*)
16.4	4(6),7(8)-dien;2α, 3α,9α-H	$C_{15}H_{20}O_2$	–	–	–	(*81*)
16.5	6(7),8(9)-dien;2α, 3α-H	$C_{15}H_{20}O_2$	oil	+20.3	–	(*94*)
16.6	6(7),8(9)-dien;2α, 3β-H	$C_{15}H_{20}O_2$	oil	+168.0	*L. vellereus*	(*94*)
16.7	3α-OH;6(7),8(9)-dien; 2α-H, Lactaro-scrobiculide B	$C_{15}H_{20}O_3$	–	–	*L. scrobiculatus*	(*11, 87*)
16.8	2β,9β-epoxy;7(8)-en; 3α,6β-H	$C_{15}H_{20}O_3$	oil	+102.0	*L. scrobiculatus*	(*9, 119*)
16.9	3α,8α-epoxy;6(7)-en; 2α,9α-H	$C_{15}H_{20}O_3$	oil	+24.6	–	(*97*)
16.10	3α,5-diOH;8-oxo;6(7)-en;2α,9α-H	$C_{15}H_{20}O_5$	–	–	–	(*84*)
16.11	3α-OH;6(7)-en;2α, 9α-H	$C_{15}H_{22}O_3$	oil	+31.0	–	(*11, 84*)
16.12	3α,8α-diOH;6(7)-en; 2α,9α-H	$C_{15}H_{22}O_4$	68–72	+9.5	–	(*84, 87, 97, 114*)
16.13	3α,8β-diOH;6(7)-en; 2α,9β-H	$C_{15}H_{22}O_4$	oil	+7.3	–	(*111*)
16.14	3α,5,8α-triOH;6(7)-en; 2α,9α-H, Lactarolide B	$C_{15}H_{22}O_5$	212–216	−3.5	*L. scrobiculatus*, *L. pallidus* *L. blennius*	(*84, 87, 114*)
16.15	8α-OAc;3(4), 6(7)-dien;2α,9α-H	$C_{17}H_{22}O_4$	oil	–	–	(*114*)

(continued on p. 114)

(continued from p. 113)

No.	Substituents, name	Formula	m.p. °C	$[\alpha]_D{}^{20}$	Source	Ref.
16.16	3α-OEt;8-oxo;6(7)-en; 2α,9α-H	$C_{17}H_{24}O_4$	144–145	+38.5	–	*(116)*
16.17	8α-OAc;6(7)-en; 2α,3α,9α-H	$C_{17}H_{24}O_4$	95	−77.7	–	*(94)*
16.18	8α-OAc;6(7);-en; 2α,3β,9α-H	$C_{17}H_{24}O_4$	119–121	+24.7	–	*(94)*
16.19	3α-OH;8α-OAc; 6(7)-en;2α,9α-H	$C_{17}H_{24}O_5$	–	+22.0	–	*(84, 87, 114)*
16.20	3α-OH;8β-OAc; 6(7)-en;2α,9β-H	$C_{17}H_{24}O_5$	185–190	+33.6	–	*(111)*
16.21	3α,5-diOH;8α-OAc; 6(7)-en;2α,9α-H	$C_{17}H_{24}O_6$	–	–	–	*(84, 87)*
16.22	8α-OH;3α-OEt; 6(7)-en;2α,9α-H	$C_{17}H_{26}O_4$	149–152	+9.1	–	*(84, 116)*
16.23	8β-OH;3α-OEt; 6(7)-en;2α,9β-H	$C_{17}H_{26}O_4$	62	+16.2	–	*(118)*
16.24	5,8α-diOH;3α-OEt; 6(7)-en;2α,9α-H, 3-O-Ethyllactarolide B	$C_{17}H_{26}O_5$	180–183	+2.1	*L. pallidus, L. blennius*	*(84)*
16.25	3α-OH;5,8α-diOAc; 6(7)-en;2α,9α-H	$C_{19}H_{26}O_7$	–	−28.4	–	*(84)*
16.26	3α-OH;5α,8α-diOAc; 6(7)-en;2α,9α-H	$C_{19}H_{26}O_7$	oil	−44.7	–	*(87)*
16.27	3α-OH;5β,8α-diOAc; 6(7)-en;2α,9α-H	$C_{19}H_{26}O_7$	–	+1.0	–	*(87)*
16.28	3α-OH;8α,15-diOAc; 6(7)-en;2α,9α-H	$C_{19}H_{26}O_7$	61–63	−12.3	–	*(87)*
16.29	3α,5-diOH;8α, 15-diOAc;6(7)-en; 2α,9α-H	$C_{19}H_{26}O_8$	–	–	–	*(87)*
16.30	3α-OEt;8α-OAc; 6(7)-en;2α,9α-H	$C_{19}H_{28}O_5$	oil	−17.0	–	*(116)*
16.31	3α-OEt;8α-OAc; 6(7)-en;2α,9β-H	$C_{19}H_{28}O_5$	oil	+48.8	–	*(118)*
16.32	3α-OH;5α,8α, 15-triOAc;6(7)-en; 2α,9α-H	$C_{21}H_{28}O_9$	oil	−10.0	–	*(87)*
16.33	3α-OH;5β,8α, 15-triOAc; 6(7)-en;2α,9α-H	$C_{21}H_{28}O_9$	oil	−8.2	–	*(87)*
16.34	3α-OEt;5,8α-diOAc; 6(7)-en;2α,9α-H	$C_{21}H_{30}O_7$	–	–	–	*(84)*

Part 17. 13-Lactaranolide Derivatives

The only example of this class is the synthetic compound **17.1**, whose formation from chrysorrheal (**10.10**) on exposure to *p*-TsOH in EtOH confirmed the position of the OH and CHO groups in the latter sesquiterpene (*9*). Characteristic ^{1}H NMR signals of structure **17.1** are the broad singlet at δ 5.90 of H-8 and the two singlets at δ 5.30 and 5.45, respectively, assigned to the H-13 proton of the two epimeric hemiacetals.

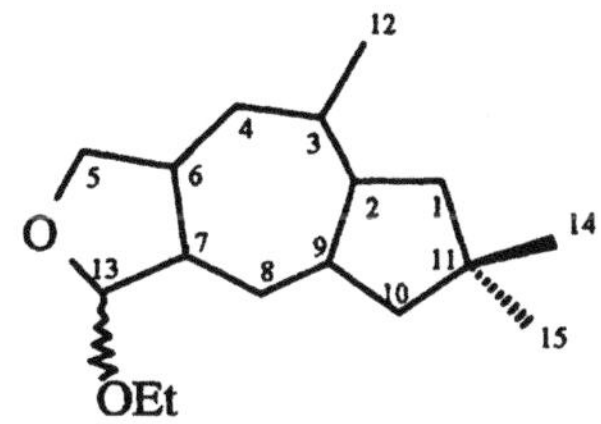

No.	Substituents, name	Formula	m.p. °C	$[\alpha]_D^{20}$	Source	Ref.
17.1	2(9),7(8)-dien;3α, 6β-H	$C_{17}H_{26}O_2$	–	–	–	(*9*)

Part 18. Furanolactarane Sesquiterpenes

Furanol **18.11** and furandiol **18.14** were the first furanolactarane sesquiterpenes isolated from mushrooms, though not from a *Lactarius* species. In fact, they were discovered in 1971 by SHIGEO NOZOE and collaborators (*65*) in extracts of the cultured mycelium of *Fomitopsis insularis*. Later, a great number of furanoid sesquiterpenes, including **18.11** and **18.14**, were isolated from several pungent *Lactarius* species. Furanolactaranes have long been regarded with some skepticism as true natural products, since the fatty acid esters of velutinal (e.g. **7.28** and **7.30**), originally present in intact fruiting bodies, are chemically labile and rapidly degraded in several reagent grade solvents, mainly in alcohols, or during chromatography, to dihydrohydroxyfurans (Part 12) and furans via cations formed from acid catalysed opening of the epoxide ring (see Scheme 15, p. 134) (*12*). Furan sesquiterpenes (Parts 18 and 19) were often isolated in large amounts from *Lactarius* species, when alcohol or acetone was used as solvent for extraction, but for the reasons stated above, they are surely artifacts. However, it is now accepted that some furans, for example diene **18.2** and furanol **18.11** in *L. scrobiculatus* and *L. chrysorrheus* (*9*), respectively, furandiols **18.14** and **18.18** in *L. piperatus* (*120*), *L. torminosus* (*120*), *L. necator* (*47*), and *L. circellatus* (*47*), furantriol (**18.19**) in *L. mitissimus* (*110*) can also be formed from velutinal esters by enzymatically assisted conversions in injured fruiting bodies, although only in small amounts. The enzymatic pathway may likely be similar to the mechanism involved in the chemical transformations of velutinals (Scheme 15, p. 134) (*9*). In conclusion, any isolation of a furanoid compound from a *Lactarius* species must be considered with some suspicion, unless appropriate procedures for extraction and chromatografic separation of individual compounds have been strictly followed (*3–9*).

Besides the structure of the highly oxidised dioxofuran **18.1**, verified by single crystal X-ray diffraction analysis (*121*), the stereostructures of other furanoids were established mainly by spectroscopic methods and simple chemical transformations. In addition, molecular mechanics (MM) calculations were carried out to determine unambiguously the relative stereochemistry of diol **18.18** (*120*) and of furoscrobiculin D (**18.13**) (*90*). For all furanolactaranes, IR bands at *ca.* 1540 and 880 cm^{-1} are diagnostic of the furan ring. Characteristic features of the ^{1}H NMR spectra are the signals of furan protons at C-5 and C-13, that usually appear in the range of δ 7.10–7.20 and are coupled, with coupling constants of *ca.* 1.5 Hz, to each other and to H-4 and H-8 (where applicable). When a free hydroxy group is attached to C-4 or C-8, the nearby furan proton is shifted downfield to *ca.* δ 7.35. A characteristic signal of 8-hydroxyfuranolactaranes is that of H-8, which usually resonates at δ 4.30–4.40 and is moved downfield to δ 4.60–4.80 when another hydroxy or alkoxy group is located at C-3 or C-4. The associated coupling constant H-8 / H-9 is *ca.* 10.5 Hz, which corresponds to a dihedral angle of *ca.* 180°, except when the formation of an intramolecular hydrogen bond across the ring induces an entirely different folding of the 7-membered ring. Thus, in furandiols **18.14** and **18.18**, $J_{8,9}$ is equal to 4.4 and 6.8 Hz, respectively (*50*, *65*, *120*). Other NMR signals of furanolactaranes follow the general pattern of the spectra of related lactaranolide sesquiterpenes.

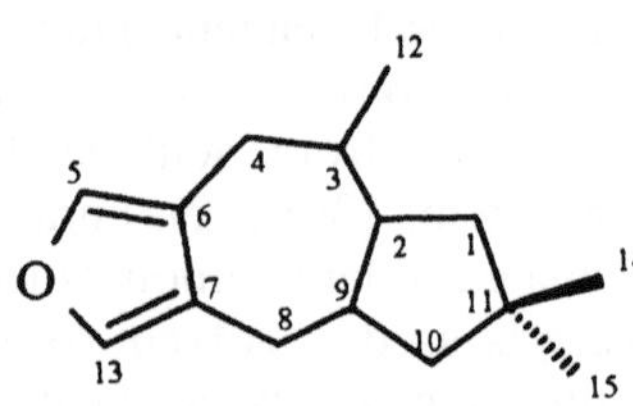

No.	Substituents, name	Formula	m.p. °C	$[\alpha]_D^{20}$	Source	Ref.
18.1	4,8-dioxo;3β-OH; 2(9)-en	$C_{15}H_{16}O_4$	100–1	+123.3	*L. vellereus*	(*93, 121*)
18.2	1(2),8(9)-dien;3α-H	$C_{15}H_{18}O$	oil	−4.2	*L. scrobiculatus*	(*9, 12, 98*)
18.3	2(3),8(9)-dien	$C_{15}H_{18}O$	oil	–	–	(*12*)
18.4	3(12),8(9)-dien; 2α-H;Pyrovellerofuran	$C_{15}H_{18}O$	oil	+189.4	–	(*122*)
18.5	2,9-epoxy;8-oxo; Furoscrobiculin A	$C_{15}H_{18}O_3$	oil	−23	*L. scrobiculatus*	(*11*)
18.6	3(4)-en;2α,9α-H	$C_{15}H_{20}O$	oil	−111.9	–	(*81*)
18.7	3α,8α-epoxy;2α,9α-H; Furanether A	$C_{15}H_{20}O_2$	oil	+29.9	*L. scrobiculatus*, *L. necator*, *Russula sardonia*	(*11, 97, 112*)
18.8	3β,8β-epoxy;2α, 9α-H; Furanether B	$C_{15}H_{20}O_2$	oil	+48.6	*L. scrobiculatus*	(*11*)
18.9	3α-OH;8(9)-en;2α-H; Furoscrobiculin B	$C_{15}H_{20}O_2$	oil	+115.1	*L. scrobiculatus*, *L. hepaticus*	(*11, 123*)
18.10	8α-OH;1(2)-en;3α, 9α-H;Furosardonin A	$C_{15}H_{20}O_2$	71–73	−14.7	*Russula sardonia*	(*98,* 112)
18.11	8α-OH;2(3)-en; 9α-H;Furanol	$C_{15}H_{20}O_2$	34–44	+123	*L. vellereus*, *L. rufus*, *L. pergamenus*, *L. scrobiculatus*, *L. necator*, *L. blennius*, *L. circellatus*, *L. chrysorrheus*, *L. helvus*	(*9, 10, 47, 49, 65, 93, 103*)
18.12	8α-OH;3(12)-en; 2α,9α-H	$C_{15}H_{20}O_2$	oil	+2.8	*L. vellereus*	(*124*)
18.13	2β,8α-diOH;3α,9α-H; Furoscrobiculin D	$C_{15}H_{22}O_3$	142–3	+10.8	*L. scrobiculatus*	(*11, 90*)
18.14	3α,8α-diOH;2α, 9α-H;Furandiol	$C_{15}H_{22}O_3$	65–66	+6.0	*L. vellereus*, *L. necator*, *L. rufus*, *L. pallidus*, *L. piperatus*, *L. scrobiculatus*, *L. torminosus*, *L. blennius*, *L. circellatus*, *L. trivialis*, *L. spinosulus*, *L. mitissimus*, *L. quietus*, *L. subdulcis*	(*11, 47, 49, 50, 56, 58, 60, 64, 65, 79, 87, 100, 101, 102, 103, 104, 109, 110, 114, 120*)
18.15	3β,8α-diOH;2α, 9α-H; 3-*epi*-Furandiol	$C_{15}H_{22}O_3$	oil	+20	*L. scrobiculatus*	(*92*)

(continued on p. 118)

(continued from p. 117)

No.	Substituents, name	Formula	m.p. °C	$[\alpha]_D^{20}$	Source	Ref.
18.16	3α,8β-diOH;2α, 9α-H; 8-*epi*-Furandiol	$C_{15}H_{22}O_3$	oil	+20	–	(*111*)
18.17	3α,8β-diOH;2α,9β-H; 8-*epi*-9-*epi*-Furandiol	$C_{15}H_{22}O_3$	140	+14.8	–	(*111, 125*)
18.18	4α,8α-diOH;2α, 3β,9α-H	$C_{15}H_{22}O_3$	oil	+4	*L. piperatus, L. necator, L. torminosus, L. circellatus*	(*47, 120*)
18.19	3α,8α,15-triOH; 2α,9α-H; Furantriol	$C_{15}H_{22}O_4$	57–58	+16.0	*L. mitissimus*	(*87, 109, 110*)
18.20	8α-OH;3α-OMe; 2α,9α-H;3-O-Methylfurandiol	$C_{16}H_{24}O_3$	65–66	+6	*L. vellereus, L. pergamenus, L. helvus*	(*10, 12*)
18.21	8α-OAc;2α,3β,9α-H	$C_{17}H_{24}O_3$	oil	–	–	(*94*)
18.22	2β-OH;8α-OAc; 3α,9α-H,	$C_{17}H_{24}O_4$	oil	–	–	(*11, 90*)
18.23	3α-OH;8α-OAc; 2α,9α-H	$C_{17}H_{24}O_4$	118–22	−22.5	–	(*50, 114*)
18.24	3β-OH;8α-OAc; 2α,9α-H	$C_{17}H_{24}O_4$	oil	–	–	(*92*)
18.25	3α-OH;8α-OEt;2α, 9α-H; Furoscrobiculin C	$C_{17}H_{26}O_3$	oil	+12.8	*L. scrobiculatus*	(*11*)
18.26	8α-OH;3α-OEt;2α, 9α-H; 3-O-Ethylfurandiol	$C_{17}H_{26}O_3$	oil	+5.8	*L. vellereus, L. helvus, L. pergamenus, L. necator, L. pallidus, L. rufus, L. quietus*	(*10, 11, 56, 79, 84, 93, 116*)
18.27	8β-OH; 3α-OEt;2α, 9β-H;3-O-Ethyl-8-*epi*-9-*epi*-Furandiol	$C_{17}H_{26}O_3$	117	+5.2	–	(*118, 126*)
18.28	4α,8α-diOAc; 2α,3β,9α-H	$C_{19}H_{26}O_5$	70–75	+56	–	(*120*)
18.29	3α-OH;8α,15-diOAc; 2α,9α-H	$C_{19}H_{26}O_6$	oil	−7.1	–	(*87, 110*)
18.30	3α-OEt;8α-OAc; 2α,9a-H	$C_{19}H_{28}O_4$	95–100	−25	–	(*116, 118*)
18.31	3α-OEt;8β-OAc; 2α,9β-H	$C_{19}H_{28}O_4$	oil	+7.5	–	(*118*)
18.32	3α,8α-diOEt;2α,9α-H	$C_{19}H_{30}O_3$	–	–	–	(*11*)

Part 19. 8,9-Secofuranolactarane Sesquiterpenes

The group of 8,9-secofuranosesquiterpenes so far comprises five compounds, two of which (**19.1** and **19.2**) were isolated from mushrooms. The remaining ones are synthetic derivatives. Lactaral (**19.1**) was also obtained together with lactardial (**14.2**) as a degradation product of the velutinal esters **7.28** and **7.30** (see Scheme 15, p. 134) (*12*, *60*, *98*). The possibility that compound **19.1** isolated from extracts of injured Russulaceae species is an artifact must therefore be considered. Lactarol **19.2** was originally obtained by reduction with KBH_4 of both **14.2** and **19.1**, or when esters **7.28** and **7.30** were degraded by adsorption on alumina (*7*), though in small amounts. However, when later isolated from a few *Lactarius* species in extracts made more than 30 minutes after grinding, lactarol was considered a true metabolite of these mushrooms (*60*). Compound **19.5** is surely an artifact formed during extraction of *L. scrobiculatus* with acetone.

With respect to the aliphatic moiety (cyclopentene ring, plus protons at C-3, C-4, and C-12), 8,9-secofuranolactaranes gave ^1H NMR spectra resembling those of 8,9-seco-5-lactaranolide sesquiterpenes (Part 14); in addition, the two protons on the furan ring exhibited the characteristic couple of signals at δ 7.20–7.36. Remarkably, when a carbonyl group is attached to C-7, as in compounds **19.1** and **19.3**, the signal of H-13 is shifted downfield to δ 7.95.

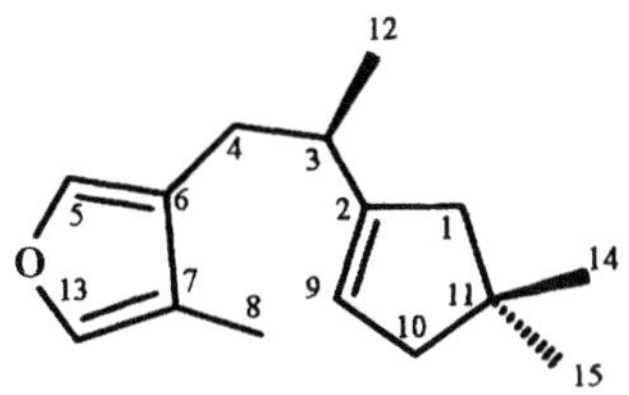

19.1-19.4

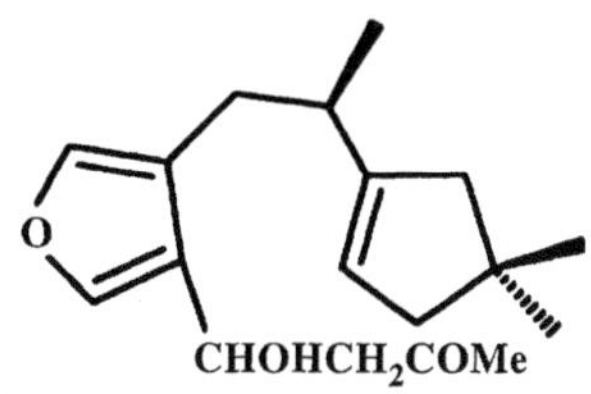

19.5

No.	Substituents, name	Formula	m.p. °C	$[\alpha]_D^{20}$	Source	Ref.
19.1	8-oxo; Lactaral	$C_{15}H_{20}O_2$	oil	−7.6	*L. vellereus*, *L. pallidus*, *L. pergamenus*, *L. scrobiculatus*	(*11*, *56*, *127*, *128*)
19.2	8-OH; Lactarol	$C_{15}H_{22}O_2$	oil	−3.5	*L. piperatus*, *L. torminosus*, *L. vellereus*, *L. necator*	(*38*, *60*, *94*, *98*, *128*)
19.3	7-COOMe	$C_{16}H_{22}O_3$	oil	−12.9	–	(*10*)
19.4	8-OTHP	$C_{20}H_{30}O_3$	oil	–	–	(*128*)
19.5	See formula	$C_{18}H_{26}O_3$	oil	–	–	(*11*)

Part 20. Rearranged Furanolactarane Sesquiterpenes

Compound **20.1** was obtained by partial hydrogenation of isovelleral (**6.2**) over a palladium catalyst, and subsequent reduction with excess KBH_4 in dioxane-water (*54*).

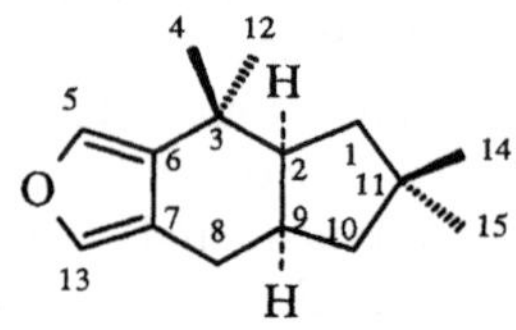

No.	Substituents, name	Formula	m.p. °C	$[\alpha]_D^{20}$	Source	Ref.
20.1	See formula	$C_{15}H_{22}O$	oil	+21.0	–	(*54*)

Part 21. Dibenzonaphthyridinone Alkaloids

As already reported in previous sections, sesquiterpenes are the most widespread *Lactarius* metabolites; however, a few species possess a particular metabolism which leads to secondary metabolites of other classes. Moreover, interesting new compounds with a different biogenesis have been isolated also from species producing large quantities of sesquiterpenes.

Interest in the considerable mutagenicity of extracts of *Lactarius necator* (syn. *L. turpis*), a mushroom often cited in this review for the occurrence of several lactarane sesquiterpenes, led to the isolation of a highly mutagenic and unstable alkaloid named necatorin (4.8 mg from 30 kg of mushrooms), for which the structure of 7-hydroxycoumaro {5,6-c}cinnolide was originally proposed (*129*). Necatorin was then shown (*130*) to be identical with necatorone (**21.2**), isolated almost at the same time by STEGLICH (*131*) as one of the pigments of the fruiting bodies of *L. necator*. The ^{1}H NMR spectrum of necatorone in DMSO-d_6 exhibited a characteristic pattern of signals between δ 6.96 and 9.03 (*132*) attributed to the protons of the dibenzonaphthyridine structure, while the ^{13}CNMR spectrum clearly demonstrated the presence in solution of a single tautomer possessing the formula **21.2**. Necatorone forms red needles {UV/Vis (MeOH): λ_{max} (log*e*): 212 (sh., 4.38), 233 (4.60), 265 (sh., 4.13), 293 (3.88), 310 (sh., 3.85), 4.31 nm (4.13)} which

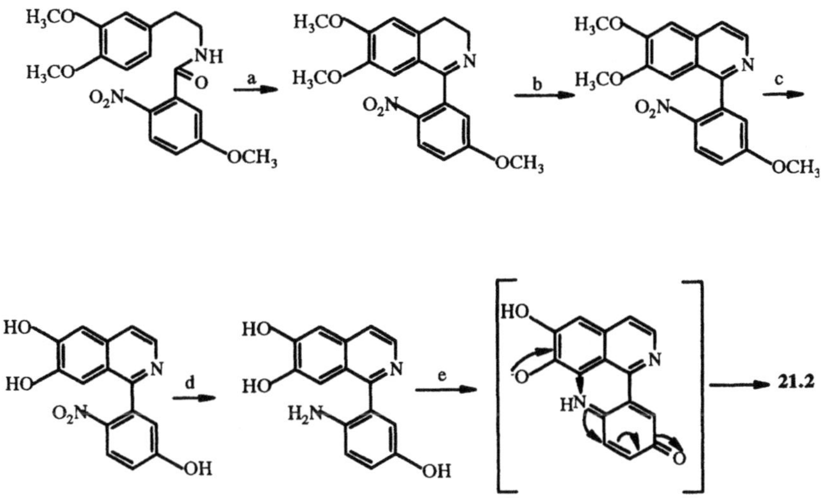

Scheme 8. a) $POCl_3$, CH_3CN, reflux, 85–93%; b) MnO_2, C_6H_6, 24 h, reflux, 90–98%; c) 48% HBr, reflux, 64%; d) H_2/Pd-C, 80–85%; e) 5% aqueous NaOH/O_2, 67%

dissolve in DMSO to produce a grass-green solution showing a strong green-yellow fluorescence. With aqueous ammonia, successive deprotonations of compound **21.2** produce blue and purple anions. The unusual 5,10- dihydroxydibenzo{de,h} {1.6}-naphthyridin-6-one structure (**21.2**) of necatorone was unambiguously confirmed by total synthesis (Scheme 8) (*130*).

Necatorone (**21.2**) was methylated by CH_2N_2 in methanol to yield the dimethylether **21.3** as the main product. More recently, two other new necatorone-type alkaloids were isolated from a MeOH extract of frozen, peeled skins from caps and stalks of *L. necator* and were identified as 4,4′-binecatorone (**21.6**) (dark brown or red crystals from MeOH) and 10-deoxy-4,4′-binecatorone (**21.5**) (dark brown crystals from MeOH) (*132*). From *L. atroviridis*, a dark-green North American species, in addition to compounds **21.2**, **21.5**, and **21.6**, 10,10′-dideoxy-4,4′-binecatorone (**21.4**) (blackish green or orange crystals from MeOH) was obtained as the main alkaloid (*132*). Pigments were separated by repeated column chromatography on Sephadex LH-20. The UV/Vis spectra of dimers were similar to those of necatorone (**21.2**), except for the maxima of dimer **21.4**, which showed a hypsochromic shift in comparison with those of **21.2**. Indicative of the 4,4′-dimeric structures **21.4**, **21.5**, and **21.6** were the lack of the proton at C-4 of necatorone (δ 6.96) and the presence, in the ^{1}H-coupled ^{13}CNMR spectra, of a singlet at δ 115 attributed to C-4,C-4′ (*132*) Interestingly, despite hindered rotation at the biaryl linkage, the dimer **21.6** exhibited no optical activity (*132*). The structures of all these alkaloids were confirmed by total synthesis (*132*). Compound **21.6** dissolves readily in DMSO to give a greenish brown colour, while **21.5** produces a green colour and **21.4** gives a dark green colour, respectively. Like necatorone (**21.2**), on reaction with alkali alkaloids **21.5** and **21.6** exhibit a purple colour reaction, while compound **21.4** gives a dove-grey colour. Compounds **21.2**, **21.5**, and **21.6** are therefore believed to be responsible for the change to a deep purple color of the dark olive-brown caps and stalks of *L. necator* on exposure to ammonia vapours.

The occurrence of the same alkaloids in *L. necator* and *L. atroviridis* indicates the close taxonomic relationship of both species. It is noteworthy that in young, light brown fruiting bodies of *L. necator* about equal amounts of pigments **21.2** and **21.6** are present, whereas in aged, dark brown specimens the ratio between these compounds becomes 5 : 95.

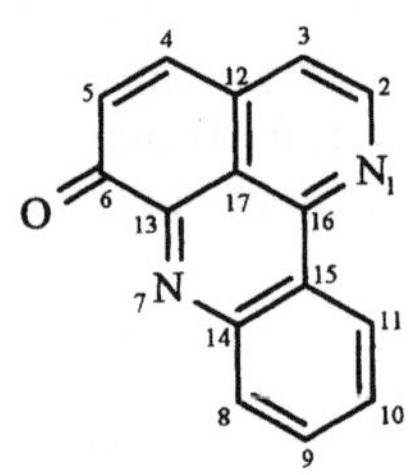

21.1-21.3

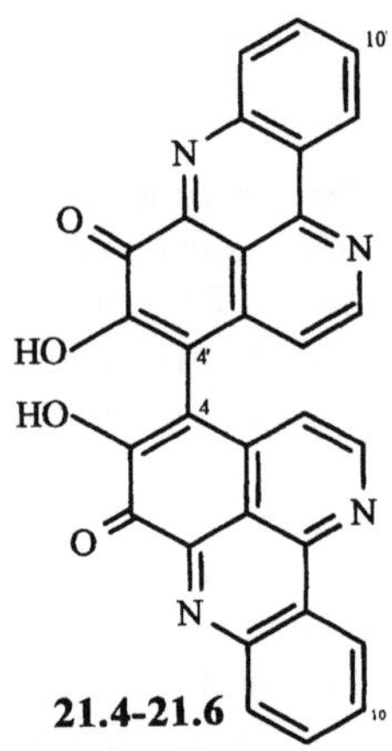

21.4-21.6

No.	Substituents, name	Formula	m.p. °C	$[\alpha]_D^{20}$	Source	Ref.
21.1	5-OH; 10-Deoxynecatorone	$C_{15}H_8N_2O_2$	–	–	–	(*132*)
21.2	5,10-diOH; Necatorone (Necatorin)	$C_{15}H_8N_2O_3$	>360	–	*L. necator*, *L. atroviridis*	(*129*, *130*, *131*)
21.3	5,10-diOMe	$C_{17}H_{12}N_2O_3$	–	–	–	(*131*)
21.4	See formula; 10,10′-Dideoxy-4, 4′-binecatorone	$C_{30}H_{14}N_4O_4$	>360	–	*L. atroviridis*	(*132*)
21.5	10′-OH; 10-Deoxy-4,4′-binecatorone	$C_{30}H_{14}N_4O_5$	>360	–	*L. necator*, *L. atroviridis*	(*132*)

(continued on p. 123)

(continued from p. 122)

No.	Substituents, name	Formula	m.p. °C	$[\alpha]_D{}^{20}$	Source	Ref.
21.6	10,10′-diOH;4,4′-Binecatorone	$C_{30}H_{14}N_4O_6$	>360	–	*L. necator*, *L. atroviridis*	(*132*)

Parts 22 and 23. Prenylated Phenols, Benzofurans, Chromanes

In search for the compounds responsible for the antimicrobial and immunosuppressive activities of *L. flavidulus*, an edible mushroom in spite of the bitter taste, four geranylphenols were isolated and named flavidulol A (**22.8**), B (**22.14**), C (**22.16**), and D (**22.13**) (*133–135*).

The structure of flavidulol A (**22.8**) is very similar to that of wigandol isolated from *Wigandia kunthii* Choisy (*136*), the former compound being the methyl ether and the latter the acetate of the same phenol. Flavidulol B (**22.14**) could be an artifact derived from flavidulol A by a Cope-type rearrangement. The structures of all the flavidulols and their acetyl derivatives (Part 22) could be determined by spectroscopic studies. In particular, NOE and ^{13}C-^{1}H-COLOC NMR techniques (*134*, *135*) established the configuration 2Z,6E of the double bonds in the geranyl moiety of compounds **22.8**, **22.13**, and **22.16** as well as the *cis* stereochemistry at C-2 and C-7 of flavidulol B (**22.14**) (*134*, *135*) and also corrected previous assignments (*133*). Catalytic hydrogenation of compound **22.8** afforded dihydro and tetrahydro derivatives, **22.11** and **22.12**, respectively, while on treatment with 2N HCl in MeOH flavidulol A (**22.8**) gave two linear tricyclic products **22.18** and **22.19** (*134*). The molecular formula of flavidulol D (**22.13**) was established by HREIMS, while a fragment ion at m/z 258 in the MS spectrum, corresponding to the loss of a stearoyl group from the molecular ion, indicated the type of ester moiety. In addition, the ^{1}H NMR spectrum of **22.13** closely resembled that of **22.12**, except for the signals assignable to the two different acyl groups.

The slightly acrid geranylgeranylhydroquinone (**22.6**) and a mixture of tasteless fatty acid esters **22.7** were isolated from the inedible species *L. lignyotus* (*137*). Clearly these phenols are biogenetically related to the flavidulols and to compound **22.1**. Methanolysis of esters **22.7** gave the hydroquinone **22.6** and a mixture of methyl esters which were identified by GC and GC-MS (*137*). Interestingly, the free hydroquinone **22.6** had previously been isolated from the sponge *Ircinia muscarum* (*138*) and as a potent contact allergen from plants of the genus *Phacelia* (*139*).

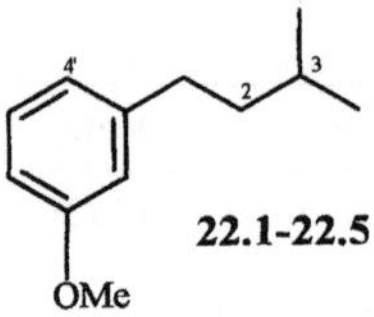

22.1-22.5

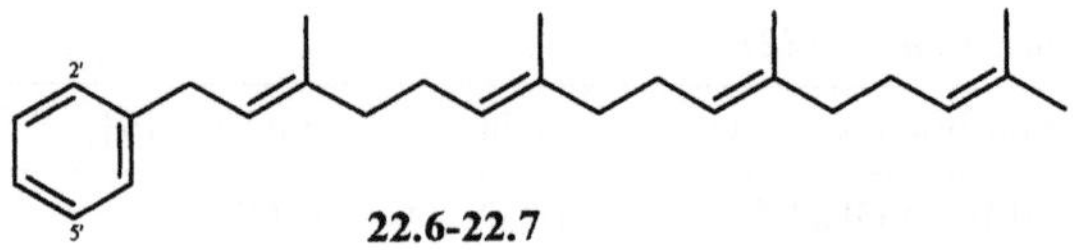

22.6-22.7

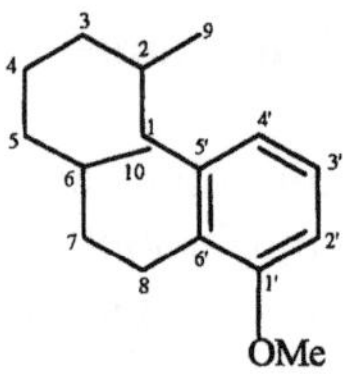

22.8-22.13

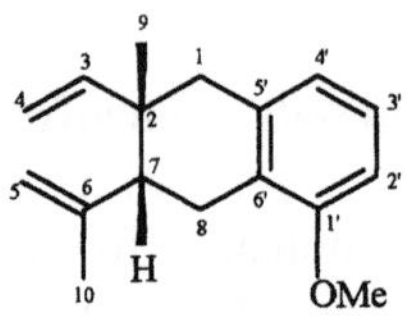

22.14-22.15

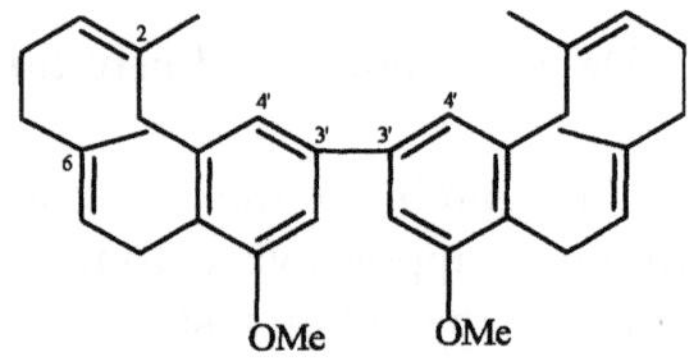

22.16-22.17

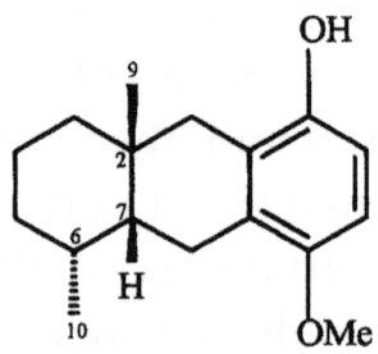

22.18-22.19

No.	Substituents, name	Formula	m.p. °C	$[\alpha]_D^{20}$	Source	Ref.
22.1	4′-OH;2(3)-en	$C_{12}H_{16}O_2$	56–59	–	*L. fuliginosus, L. picinus*	(*140*)
22.2	2,3-epoxy;4′-OH	$C_{12}H_{16}O_3$	oil	–	–	(*140*)
22.3	4′-[2-OH-3-(3-Me-2-butenyl)-5-OMe-phenoxy];2(3)-en	$C_{24}H_{30}O_4$	48–50	–	*L. fuliginosus*	(*140*)
22.4	4′-[2-OAc-3-(3-Me-2-butenyl)-5-OMe-phenoxy];2(3)-en	$C_{26}H_{32}O_5$	–	–	–	(*140*)
22.5	4′-stearoyloxy; 2(3)-en	$C_{30}H_{50}O_3$	31–32	–	*L. fuliginosus, L. picinus*	(*140*)
22.6	2′,5′-diOH	$C_{26}H_{38}O_2$	–	–	*L. lignyotus*	(*137*)
22.7	2′,5′-diOAcyl*		–	–	*L. lignyotus*	(*137*)
22.8	4′-OH;Z-2(3), E-6(7)-dien; Flavidulol A	$C_{17}H_{22}O_2$	oil	–	*L. flavidulus*	(*133, 134, 135*)
22.9	4′-OH;Z-2(3)-en	$C_{17}H_{24}O_2$	128–9	–	–	(*134*)
22.10	4′-OH	$C_{17}H_{26}O_2$	124–5	–	–	(*134*)
22.11	4′-OMe;Z-2(3), E-6(7)-dien	$C_{18}H_{24}O_2$	oil	–	–	(*133, 134, 135*)

(continued on p. 125)

(continued from p. 124)

No.	Substituents, name	Formula	m.p. °C	$[\alpha]_D{}^{20}$	Source	Ref.
22.12	4′-OAc;Z-2(3), E-6(7)-dien	$C_{19}H_{24}O_3$	111–2	–	–	(*133*, *134*, *135*)
22.13	4′-stearoyloxy; Z-2(3), E-6(7)-dien; Flavidulol D	$C_{35}H_{56}O_3$	40–41	–	*L. flavidulus*	(*135*)
22.14	4′-OH;Flavidulol B	$C_{17}H_{22}O_2$	oil	–	*L. flavidulus*	(*133*, *134*, *135*)
22.15	4′-OAc	$C_{19}H_{24}O_3$	62–64	–	–	(*133*)
22.16	4,4′-diOH; Flavidulol C	$C_{34}H_{42}O_4$	185–6	–	*L. flavidulus*	(*133*, *134*, *135*)
22.17	4,4′-diOAc	$C_{38}H_{46}O_6$	175–7	–	–	(*133*)
22.18	6β-OH	$C_{17}H_{24}O_3$	190–2	–	–	(*134*)
22.19	6β-OMe	$C_{18}H_{26}O_3$	156–7	–	–	(*134*)

* Mixture of esters of the following acids: miristic, pentadecanoic, palmitoleic, palmitic, linoleic, oleic, and stearic acid (*137*)

Separation of the EtOAc extracts of *L. fuliginosus* and *L. picinus* by silica gel column chromatography allowed isolation of the first examples of benzofuran and chromene derivatives among the constituents of *Lactarius* species. These compounds are revealed on TLC plates by the sulpho-vanillin reagent as characteristic green spots (*140*), (*141*) (Parts 22, 23).

The structures of **22.3**, **23.2**, **23.6**, and **23.10–23.14** were elucidated by spectroscopic methods. In particular, the structures of chromenes **23.11** and **23.14** were established by NOE experiments and biosynthetic considerations (*140*).

6-Methoxy-2,2-dimethylchromene (**23.6**) and benzofuran **23.2** were also synthesized by alkylation of 4-methoxyphenol to 2-(3-methyl-2-butenyl)-4-methoxyphenol (**22.1**) followed by acid catalyzed cyclization of the corresponding epoxide **22.2** to a mixture of isomeric compounds **23.3** and **23.7**. Dehydration of alcohol **23.7** with *p*-TsOH gave **23.6**, while NBS dehydrogenation of **23.3** afforded **23.2** (*140*).

It is worth noting that only one compound, the tasteless stearate **22.5** could be isolated from young intact fruiting bodies of *L. fuliginosus* and *L. picinus* extracted in the cold (*140*). On the other hand, lipases contained in injured fruiting bodies of the same species rapidly hydrolyzed **22.5** to acrid phenol **22.1**. The ester **22.5** is thus the biogenetic precursor not only of the two cyclization products **23.2** and

23.6 but also of **22.3** and **23.10–23.13**, which can be considered dimerization products of **22.1**.

Oxidative dimerizations of phenolic compounds occur in nature by one-electron transfer C–C and C–O couplings, which are catalyzed by phenol oxidase enzymes. Reactions of the same kind probably give rise to colored materials from colorless precursors during the reddening of damaged mushrooms of the Section Plinthogali. Indeed, an intense red color developed when synthetic phenol **22.1** was added to a mush of *L. fuliginosus* from which the original metabolites had been extracted with CH_2Cl_2. The structures of the red pigments are still unknown as they remained irreversibly adsorbed on the top of chromatographic columns. However, is to be noted that this experiment afforded the same mixture of chromenes and benzofurans as originally isolated from damaged fresh fruiting bodies.

The oxidative dimerization of phenol **22.1** was simulated *in vitro*. Exposure of this compound to the complex $Cu(NO_3)_2$-pyridine gave rise to dimers **23.10** and **23.11** by a C–C coupling reaction. Compound **23.11** could be cyclodehydrogenated to **23.10** by reaction with DDQ. On the other hand, exposure of phenol **22.1** to $K_3Fe(CN)_6$ gave the product (**22.3**) of a C–O coupling, which was then transformed into **23.12** by DDQ cyclodehydrogenation (*140*) (Scheme 9).

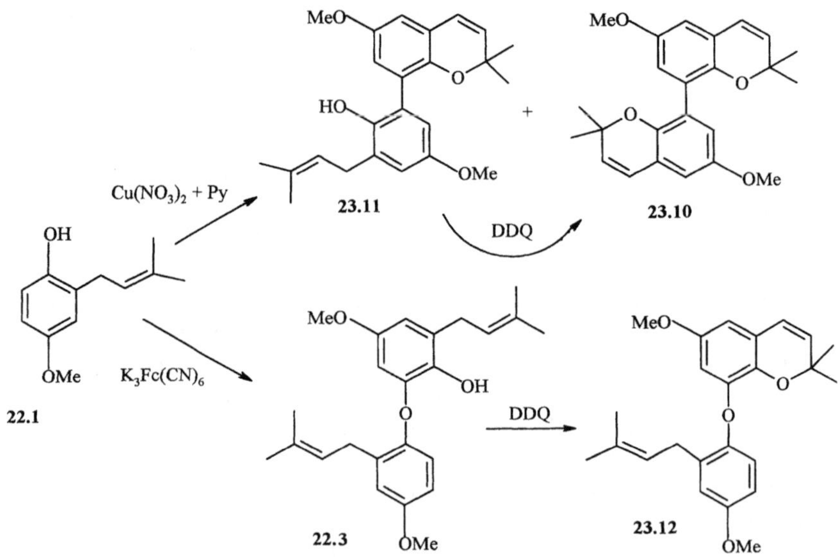

Scheme 9

In addition to the above compounds, a natural trimer (**23.14**) of phenol **22.1** was isolated from *L. fuliginosus*, and the simple 2,2-dimethylchromene (**23.4**) was identified in an extract of *L. picinus* by GC and GC–MS analysis (*140, 142*).

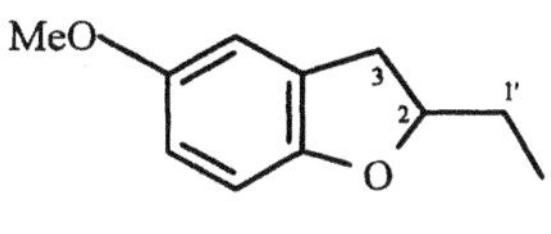

23.1-23.3

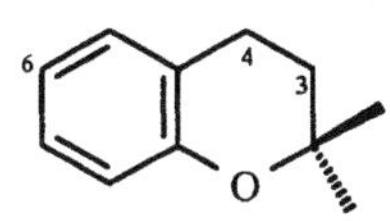

23.4-23.9

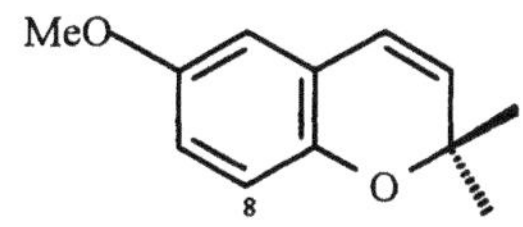

23.10-23.14

No.	Substituents, name	Formula	m.p. °C	$[\alpha]_D^{20}$	Source	Ref.
23.1	1′-oxo;2(3)-en	$C_{11}H_{10}O_3$	74–85	–	–	(*140*)
23.2	1′-OH;1′-Me;2(3)-en	$C_{12}H_{14}O_3$	44–45	–	*L. picinus*	(*140*)
23.3	1′-OH;1′Me	$C_{13}H_{16}O_3$	–	–	–	(*140*)
23.4	3(4)-en	$C_{11}H_{12}O_1$	oil	–	*L. picinus*	(*140, 141*)
23.5	4-oxo;6-CHO; Lactarochromal	$C_{12}H_{12}O_3$	87–91	–	*L. deliciosus*	(*29*)
23.6	6-OMe;3(4)-en	$C_{12}H_{14}O_2$	–	–	*L. fuliginosus*, *L. picinus*	(*140, 141*)
23.7	3-OH;6-OMe	$C_{12}H_{16}O_3$	112–115	–	–	(*140*)
23.8	6-COOMe;3(4)-en; Anofinic acid	$C_{13}H_{14}O_3$	150–156	–	*L. deliciosus*	(*29*)
23.9	3-OAc;6-OMe	$C_{14}H_{18}O_4$	oil	–	–	(*140*)
23.10	8-(2,2-diMe-6-OMe-2-H-chromen-8-yl)	$C_{24}H_{26}O_4$	–	–	*L. fuliginosus*, *L. picinus*	(*140, 141*)
23.11	8-[2-OH-3-(3-Me-2-butenyl)-5-OMe-phenyl]	$C_{24}H_{28}O_4$	–	–	*L. fuliginosus*, *L. picinus*	(*140*)
23.12	8-[2-(3-Me-2-butenyl)-4-OMe-phenoxy]	$C_{24}H_{28}O_4$	–	–	*L. fuliginosus*, *L. picinus*	(*140*)
23.13	8-[2-OH-4-OMe-6-(3-Me-2-butenyl) phenoxy]*	$C_{24}H_{28}O_5$	–	–	*L. fuliginosus*	(*140*)
23.14	8-{2-[2-(3-Me-2-butenyl)-4-OMe-phenoxy]-4-OMe-6-(3-Me-2-butenyl)phenoxy}	$C_{36}H_{42}O_6$	–	–	*L. fuliginosus*	(*140*)

* Alternatively, compound **23.11** may have the following substituent: 8-[2-OH-3-(3-Me-2-butenyl)-5-OMe-phenoxyl (*140*)

The isolation of such strictly biosynthetically related aromatic compounds from *L. lignyotus*, *L. picinus*, and *L. fuliginosus* which belong to the same section, Plinthogali (Bull.)Sing. of Bon's subdivision of the genus *Lactarius* (*2*) is taxonomically relevant. Outside this section, other chromane derivatives have been isolated only from *L. deliciosus*, which belongs to section Dapetes; however, these metabolites were not produced by wild mushrooms but when the fungus was grown in liquid cultures (*29*). Anofinic acid (**23.8**) and a new chroman-4-one lactarochromal (**23.5**) were thus obtained, along with a new indole isolated as its N,O-diacetyl derivative **24.2**, known cyclic dipeptides, ergosterol, and a mixture of fatty acids.

Part 24. Miscellaneous Compounds

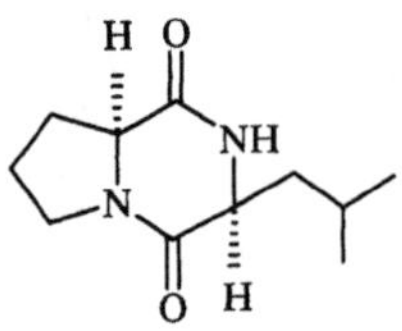

24.1

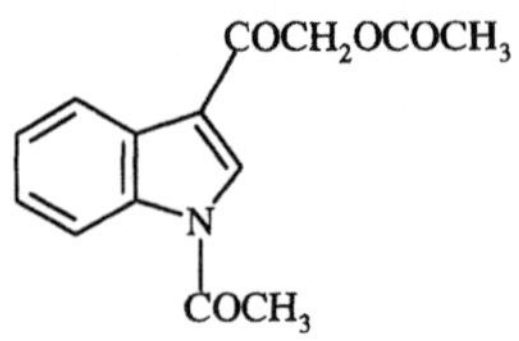

24.2

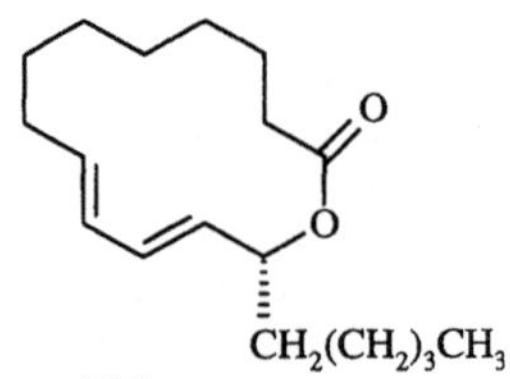

24.3

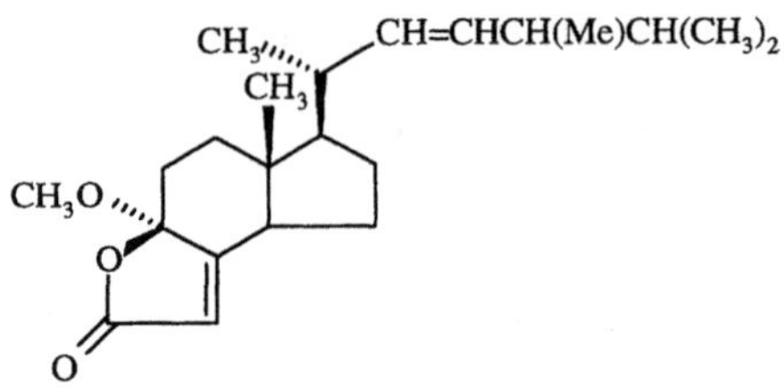

24.4

No.	Substituents, name	Formula	m.p. °C	$[\alpha]_D^{20}$	Source	Ref.
24.1	See formula	$C_{11}H_{18}N_2O_2$	155	−142	*L. vellereus*	(*67*)
24.2	See formula	$C_{14}H_{13}NO_4$	oil	–	–	(*29*)
24.3	See formula; Lactariolide	$C_{18}H_{30}O_2$	–	–	*L. subvellereus*	(*143*)
24.4	See formula; Volemolide	$C_{22}H_{34}O_3$	–	–	*L. volemus*	(*144*)

4. Chemistry of Sesquiterpenes of *Lactarius*

The discussion of the chemistry of sesquiterpenes of *Lactarius* will comprise the following aspects:

a. Interconversions and reactions (Part 25).
b. Total syntheses (Part 26).

In the following schemes some carbon skeletons are drawn in a manner different from that used so far in this review, because they have been taken directly from the cited articles and the present authors did not wish to introduce changes.

Part 25. Interconversions and Reactions

Because of space limitations, standard reactions of *Lactarius* sesquiterpenes will not be described when usual reagents were employed and the outcome of the reactions was unexceptional. It must be said, however, that preparation of simple derivatives of sesquiterpenes of *Lactarius* was often an essential part of the work of structure elucidation. In several instances, for example, crystalline compounds suitable for X-ray analysis were produced. In addition, many semisynthetic derivatives often showed better biological activities than the natural products. Important synthetic derivatives are included in Parts 1–20.

The very labile sesquiterpene velutinal (**7.12**) was prepared in 77% yield from isovellerol (**7.11**), by employing a mild variety of the Sharpless epoxidation procedure. Isovellerol (**7.11**) could be prepared, by partial reduction of isovelleral (**6.2**) in 50% yield (*72*) (Scheme 10).

It is worthy noting that while oxidation of the tetracyclic ring system **7.11** provided the desired β-epoxide **7.12** (see also compound **7.3**), Heathcock (*145*) found that epoxidation of the tricyclic hydrindane derivative **6.4** (Scheme 11) occurred from the α-face either with mCPBA or with $VO(acac)_2$ and ${}^tBuOOH$, or with dimethyldioxirane. It has been suggested that in the case of **6.4** a stereoelectronic effect is responsible for the observed α-stereoselectivity. Regioselective oxidation of diol **25.1** then provided **25.2**, a diastereomer of (±)velutinal.

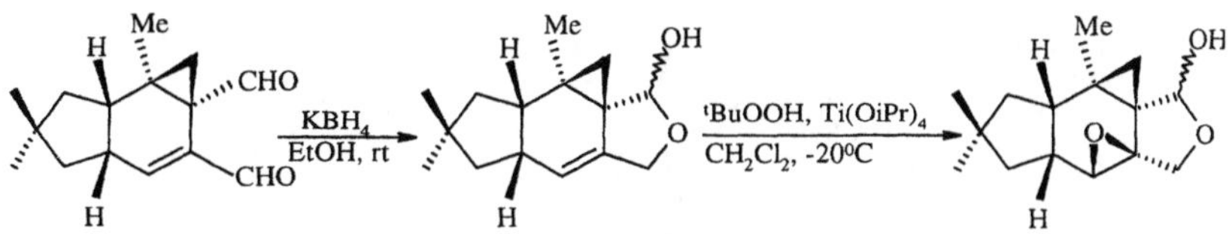

6.2 **7.11** **7.12**

Scheme 10

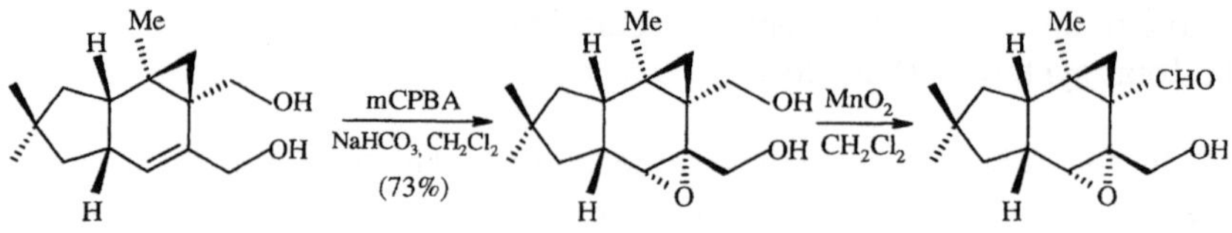

6.4 25.1 25.2

Scheme 11

The reverse conversion of velutinal (**7.12**) to isovelleral (**6.2**), was formally accomplished in two steps, starting from the methyl acetal **7.17** (*17*) (Scheme 12).

On exposure to Red-Al, stearoylvelutinal (**7.30**) was smoothly reduced and rearranged in a single step to lactaropallidine (**6.5**), another marasmane sesquiterpene (*56*). This epoxide-ketone rearrangement was probably initiated by an aluminium species which coordinated to the oxygen of the oxirane ring and promoted the {1.2} suprafacial migration of the hydrogen of the epoxide to give the corresponding ketone.

On heating in refluxing mesitylene for 0.5 h isovelleral (**6.2**) underwent a reversible rearrangement (Scheme 13) affording diastereoisomer **6.10** (ratio **6.10** : **6.2** *ca.* 3 : 2) with an inverted orientation of the cyclopropane ring (*59*, *146*). This process involved an intramolecular ene reaction through a bicyclic enol intermediate **25.3** which was trapped as the corresponding (E)-O-silyl ether. In the presence of excess D_2O, deuterium is incorporated quantitatively into the C-12 methyl groups of **6.2** and **6.10**. The thus labelled isovelleral (**6.1**) was later employed for studies on the bioconversion of sesquiterpenes in *Lactarius* species (*17*). In the presence of a weak acid or base, an equilibrium is set between isovelleral (**6.2**), **6.10**, and the hydroazulenic dialdehydes **10.3** and **10.4**. Under more vigorous conditions, i.e. higher temperatures or higher concentration of acetic acid, pyrovellerofuran **18.4** was formed, thus allowing a chemical interrelation between the marasmane and (furano) lactarane skeletons (Scheme 13) (*122*).

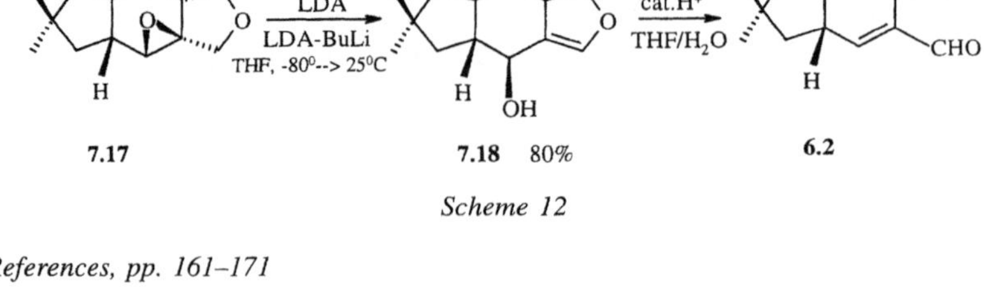

7.17 7.18 80% 6.2

Scheme 12

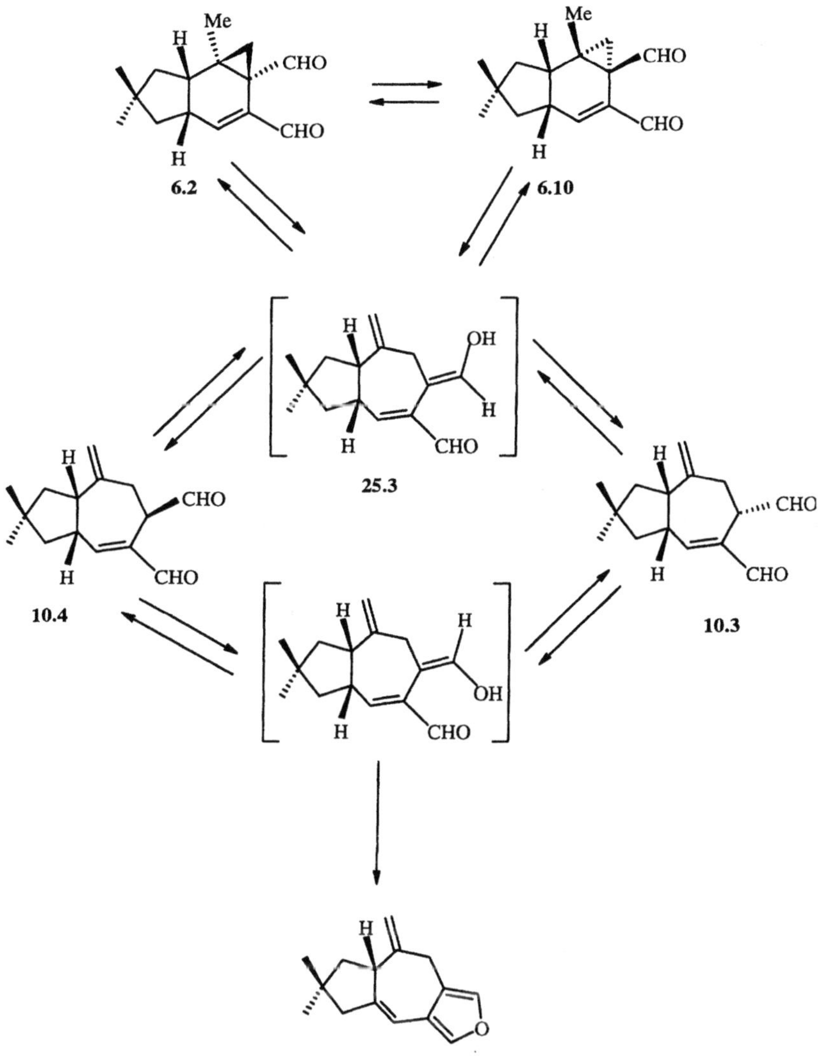

Scheme 13

This novel thermal rearrangement was also extended to the conversion of 9-hydroxyisovelleral (**6.3**) into its stereoisomer (**6.11**), and of isovellerol (**7.11**) into its isomarasmane stereoisomer (*59*). Other interesting thermal rearrangements were exhibited by vellerolactone (**11.6**) and the isomeric lactone **16.4** (*81*, *96*). In both compounds, in their preferred conformation with the Me-12 group in a pseudoequatorial position, H-3 is in a favourable position for a 1,5-sigmatropic suprafacial hydride shift, thus affording the corresponding pyrolactone, **11.7** and **16.3**, respectively, on heating in toluene at reflux (Scheme 14).

During structural studies on isovelleral (**6.2**) the cyclopropane ring was opened by partial hydrogenation over a palladium catalyst (1 equiv., H_2 uptake) and subsequent reduction with excess KBH_4 in dioxane-H_2O afforded the furan derivative **20.1**. This result proved unequivocally that the two carbonyl groups in isovelleral (**6.2**) were vicinal while the formation of a new methyl group showed that the cyclopropane hydrogens were geminal (*54*).

Degradation reactions of a few *Lactarius* sesquiterpenes were often observed. For example, velleral (**10.6**) readily decomposed on attempted preparative chromatography on Al_2O_3 (*7*) and it was slowly oxidized to 9-hydroxyvelleral (**10.7**) when a hexane solution was kept at r.t. for two weeks or at −30° for months (*7*). Under the same conditions, isovelleral (**6.2**) was oxidized to 9-hydroxyisovelleral (**6.3**) (*7*). This oxidation also occurred when isovelleral was adsorbed on Al_2O_3 for 5 h in daylight. During chromatography on silica gel, *epi*-piperdial (**10.12**) was easily converted to a mixture of velleral (**10.6**) and the apparently more stable epimer piperdial (**10.11**) (*47*). Like velleral, both piperdial and *epi*-piperdial rapidly decomposed within few seconds when chromatography on Al_2O_3 was attempted (*38*).

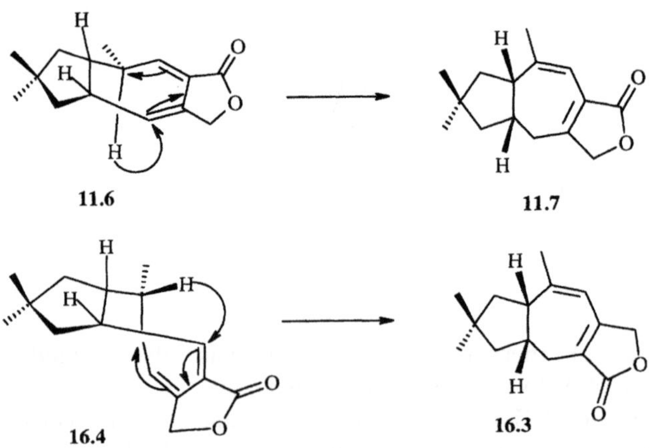

Scheme 14

References, pp. 161–171

Vellerol (**10.9**) as well as isovellerol (**7.11**), piperalol (**10.16**), and *epi*-piperalol (**10.17**) readily dimerized in reagent grade solvents where traces of acid are probably present (*7*), (*38*).

Free velutinal (**7.12**), its esters **7.28** and **7.30**, and methyl acetal **7.17** are labile compounds and on adsorption on silica gel yielded several furanolactarane and secofuranolactarane sesquiterpenes, identical with compounds previously isolated from different *Lactarius* species (see Parts 18, 19). Fast degradation took place also on dissolving velutinals in wet acetone or in reagent grade alcohols as well as under other conditions where traces of acids were probably present (*8*, *68*, *124*). Furthermore, when **7.12** was dissolved in aqueous methanol (*145*) product distribution was found to vary with acid concentration. Degradation of stearoylvelutinal (**7.30**) by adsorption on Al_2O_3 yielded, in addition to the furans, significant amounts of isovellerol **7.11** and lactarol **19.2** (*7*). The furanoid sesquiterpenes were believed to be formed via intermediate dihydrofurans, many of which, in absence of an acid catalysis, were stable enough to be isolated (*12*). The formation *in vitro* of dihydrofurans and furans from velutinal derivatives could be explained by a general mechanism *via* carbocation rearrangements (Scheme 15) which was corroborated by the stereochemistry of the sesquiterpenes formed that way (*12*, *68*). Moreover, this mechanism might mimic the enzymatic conversion of velutinal esters to the furanolactaranes and secofuranolactaranes isolated from injured mushrooms (*9*).

On exposure to protic acids synthetic lactone **25.4** (Scheme 16) underwent a ring-expansion reaction to lactaranes similar to that observed for velutinal (**7.12**), although much more drastic conditions were required (*145*). In fact, subjection of **25.4** to catalytic sulphuric acid in THF provided (±)-deconjugated anhydrolactarorufin A (**11.11**) which, within 3 days on standing, was converted into the conjugated isomer (±**11.12**). On the other hand, reaction of compound **25.4** with cat. H_2SO_4 in THF/H_2O (1 : 1 mixture) gave (+)**11.11** and (+)**11.30** (Scheme 16).

During an attempted conversion of blennin A (**11.26**) into vellerolactone (**11.6**), it was observed that upon exposure to DBU the lactarane skeleton of methanesulphonate **11.48** rearranged to the tetracyclic lactone **13.1**. The suggested mechanism of this reaction is shown in Scheme 17 (*105*).

$NaBH_4$ reduction of compound **11.25** (*104*) afforded the rearranged lactone **13.2**, which was formed by lactone ring opening of the initially formed 3-deoxy-6β,7β-dihydro-8-epilactarorufin A, followed by ring closure onto the 8α-hydroxy group.

There are many examples of chemical correlations between lactarane sequiterpenes of the different classes. These experiments have often been

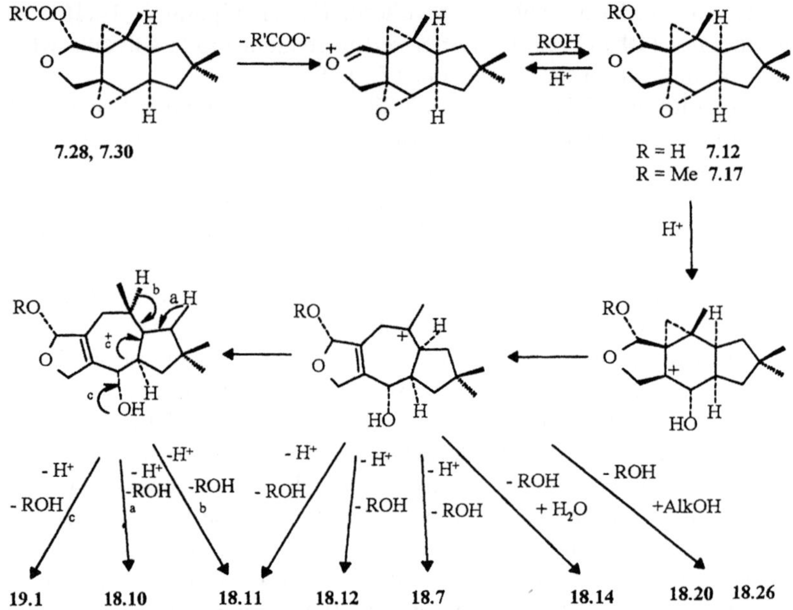

Degradation mechanism of velutinal derivatives

Scheme 15

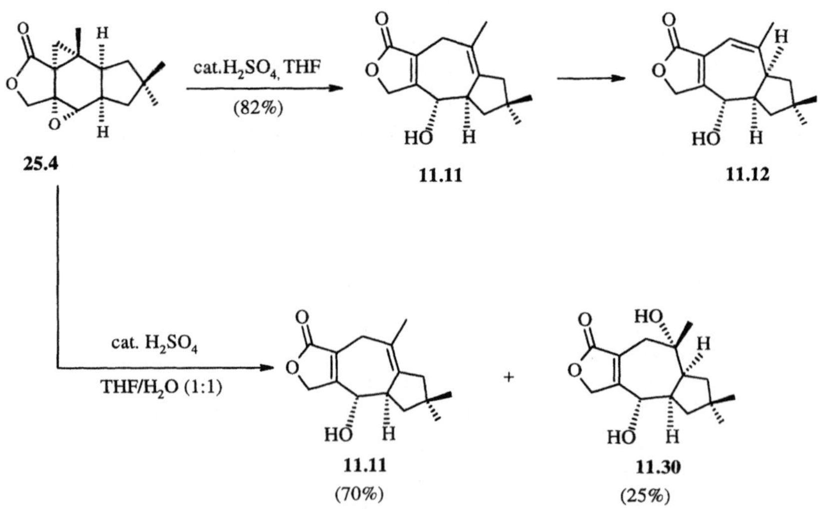

Scheme 16

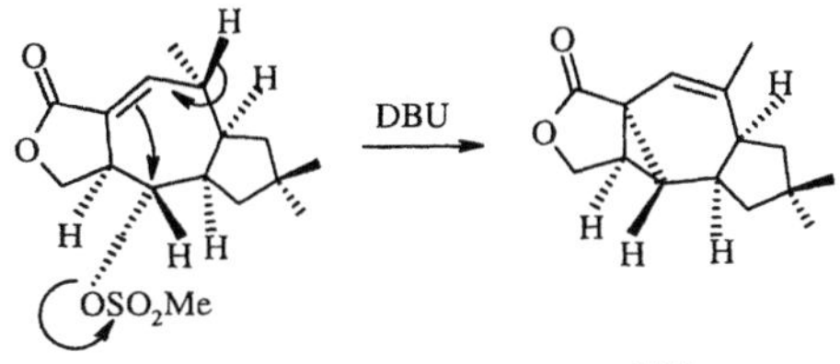

11.48 13.1

Scheme 17

conclusive for establishing not only the overall stereostructure, but also important structural details of intercorrelated compounds, namely the position of the lactone carbonyl in lactaranolides which might be either at C-5 or at C-13, and the configuration of the methyl group at C-3, which might be either *cis* or *trans* to 2-H and 9-H. In most cases standard reagents and procedures have been employed and outcomes of the reactions were unexceptional. Interesting examples are reported in Schemes 18, 20, and 22. Remarkable among the reactions reported in

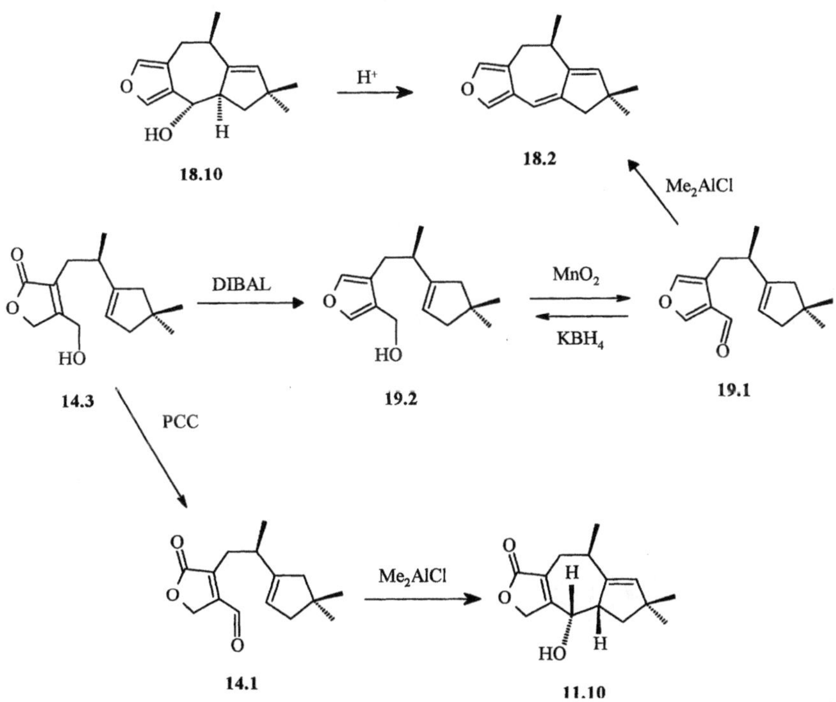

18.10 18.2 14.3 19.2 19.1 14.1 11.10

Scheme 18

Scheme 18 is the Me_2AlCl catalyzed ene cyclisation of 8,9-seco-furanolactarane and 8,9-seco-5-lactaranolide sesquiterpenes exemplified by compound **14.1** to the corresponding lactarane sesquiterpenes such as **11.10** (*98*). Lactone (**11.10**) exhibited the unusual *cis* configuration between H-8 and H-9. The emergence of this stereorelationship could be anticipated by examining the Dreiding models of the two possible transition states **25.5** and **25.6** (Scheme 19). In fact, unfavourable steric interactions developing between the C-3 methyl group and the bulky >C=O···Al– complex are minimized in the transition state **25.6**, which eventually collapses to lactone **11.10** (*98*).

Oxidative hydroboration (*91*) of lactone **11.49** afforded a mixture of four lactarorufins (Scheme 20), in which diols **11.30** and **11.33**, arising from α attack of diborane, largely predominated (more than 90%). The same type of stereoselectivity was observed for other addition reactions, *i.e.* epoxidation, osmylation, hydrogenation (*115*), (*104*), (*43*) to 2,9, 3,4- or 6,7-double bond of lactaranolides and marasmanes. Apparently, the tricyclic structures of these substrates provided enough conformational and steric bias to direct approach of reagents from the same side as the bridgehead protons H-2 and H-9. However, when the double bond was located in a different position, exceptions were observed (*98*).

As expected, Pd catalysed hydrogenation of the C_3-C_4 double bond of lactone **16.15** (Scheme 7) afforded the dihydroderivative **16.17** in which the C-3 methyl group was *trans* to H-2 (*94*). Comparison of the NMR data of compound **16.5**, readily prepared from **16.17**, with those of natural lactone **16.6** (Scheme 6) definitely proved the stereostructure of the latter sesquiterpene (*94*). Compound **16.6** was also synthesized from 5-lactaranolide **11.28** according to the reaction sequence shown in Scheme 6, which is a nice example of a general strategy for moving the carbonyl group of lactaranolides from C-5 to C-13 (*94*).

3,8-Ether formation from the corresponding 3,8-furanolactarane or 3,8-lactaranolide diol was performed using different procedures. Thus

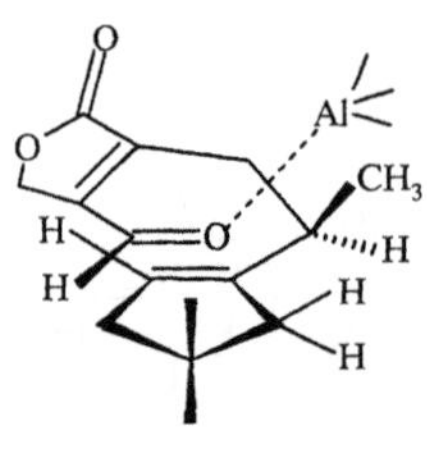

25.5

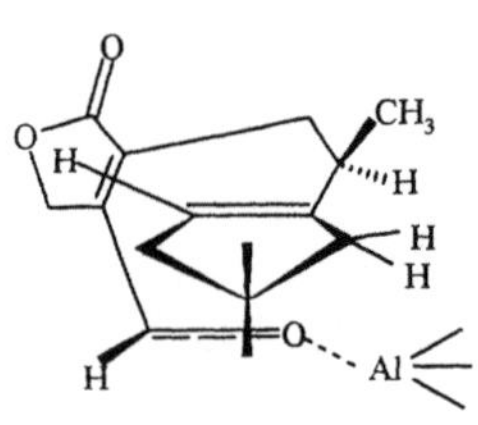

25.6

Scheme 19

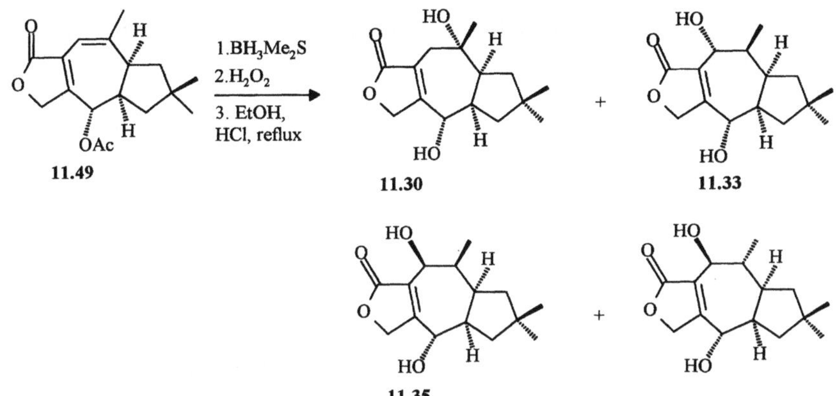

11.49 11.30 11.33 11.35 11.36

Scheme 20

lactarorufin A (**11.30**) gave the 3,8-internal ether **11.14** by dehydration with MsCl in pyridine (*83*), while furandiol **18.14** and 5-deoxylactarolide B (**16.12**) afforded **18.7** and **16.9** (Scheme 21), respectively, albeit in low yield, by acid catalyzed azeotropic removal of H_2O (*97*).

On the other hand, attempted conversion of furanol **18.12** to the corresponding 8-bromide with CBr_4 and Ph_3P gave, instead, furanether A (**18.7**) and pyrovellerofuran (**18.4**) as main products (*124*). Several interconversions of lactaranolides to the corresponding furanosesquiterpenes were readily achieved by reduction of the γ-butenolide carbonyl group with DIBAL, followed by dehydration and aromatisation of intermediate lactol with aqueous acid. Some representative examples are shown in Schemes 6 and 18. The reverse transformation of a furan to a butenolide ring has also been carried out; either NBS in aqueous dioxane (*87*, *98*) (Wiesner procedure) (*147*) or MCPBA in CH_2Cl_2 (*87*, *93*, *94*, *109*, *110*, *111*, *114*, *116*) was used as oxidant. It was observed that in the oxidation of the furan ring the directing effects of neighbouring oxygenated groups were generally unpredictable and, therefore, variable

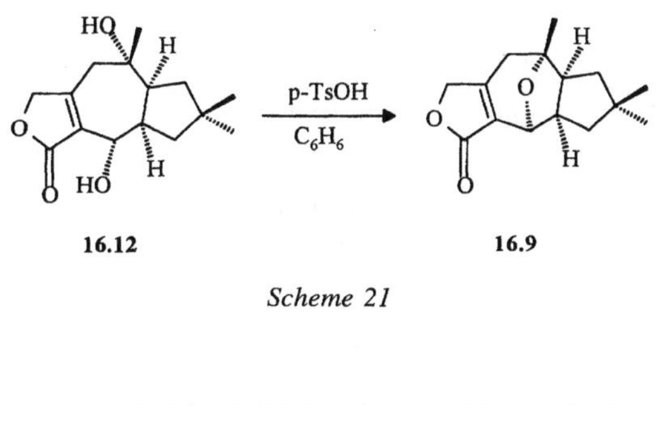

16.12 16.9

Scheme 21

mixtures of regioisomeric γ-hydroxybutenolides were obtained. Moreover, the two oxidants often showed opposite regiochemistry. After chromatographic separation, each regioisomeric lactol smoothly afforded the corresponding C-5 or C-13 lactone upon exposure to $NaBH_4$. Some representative examples are shown in Schemes 6 and 22.

All lactarane sesquiterpenes isolated so far from mushroom extracts have a *cis* fusion between the seven and five membered rings, which arises from the same precursors, i.e. velutinal esters **7.28** and **7.30** (Part 7), where such stereochemistry is created along the biosynthetic pathway. However, there are some synthetic *trans* fused lactarane sesquiterpenes. The first examples reported in the literature were the two lactones **11.23** and **11.24** (Part 11) with 2-*epi* (2βH,9αH) stereochemistry, which were obtained by catalytic hydrogenation of lactones **11.9** and **11.11**, respectively (*98*). Interesting 9-*epi*- (2αH,9βH) lactarane sesquiterpenes were synthetised starting from furandiol **18.14** (Scheme 23) (*111*). In fact, when **18.14** was submitted to the Collins reagent (chromium trioxide pyridine complex) followed by reduction with $NaBH_4$, 8-*epi*-9-*epi*-furandiol **18.17** (X-ray structure) (*125*) was produced along with the expected furandiol (**18.14**) and the 8-*epi*-stereoisomer **18.16**. This result was explained by assuming that under the basic oxidation conditions, the initially formed ketone **25.7** readily epimerized to 9-*epi* **25.9** through enolate **25.8** (Scheme 23). Therefore, subsequent reduction of **25.7** gave **18.14** and **18.16**, while **25.9** afforded **18.17**. Indeed, when the intermediate ketone was not isolated but was allowed to equilibrate in a methanolic solution of NaOH prior to

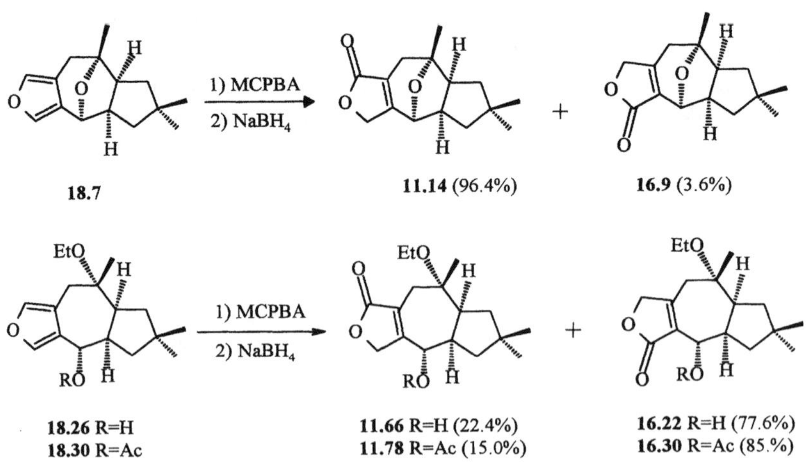

Scheme 22

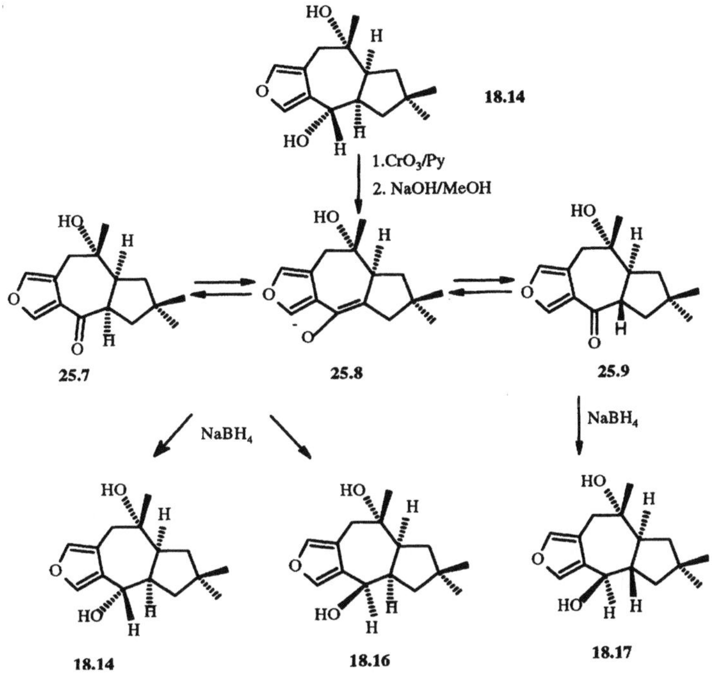

Scheme 23

reduction, compound **18.17** was obtained in almost 95% yield (*111*). It was concluded that this result reflected the higher thermodynamic stability of *trans*-fused ketone **25.9** with respect to *cis*-fused **25.7**, in agreement with the minimized molecular energy calculated (PCMODEL-4 program) for the two ketones which favoured **25.9**.

Part 26. Total Syntheses

In this chapter, we review the literature on total syntheses of *Lactarius* sesquiterpenes. For space limitations we have limited our comments to the most intriguing synthetic steps. Moreover, we have not included incomplete synthetic approaches although they often rely on imaginative novel chemistry. The chapter has been divided into sections following the division of the general part.

Drimane Sesquiterpenes

Among drimane sesquiterpenes of *Lactarius*, only uvidin C (**3.9**) has been chosen so far as a synthetic target. Two syntheses of this compound have been reported as a racemate by ZIEGLER *et al.* (*148*), and as the natural (−)-enantiomer by CORTES and colleagues (*149*).

Both approaches are based on the stereoselective epoxidation of allylic alcohol **26.10** with mCPBA. Compound **26.10** was obtained from the known (±)methyl bicyclofarnesate **26.1** and from (−)-drimenol (**3.7**), respectively, by the reaction sequences shown in Schemes 24 and 25, respectively.

During these studies, ZIEGLER made a few interesting observations on the stereoselectivity of hydroxyl-directed epoxidation of allylic alcohols **26.10**, **26.9** and the corresponding C-6 epimer **26.7**, which may be of general interest. Thus, while **26.10** was epoxidized exclusively from the β-face with either m-CPBA in CH_2Cl_2 or $VO(acac)_2$/TBHP (3 mole %, CH_2Cl_2), epoxidation of the α-alcohol **26.7** with both reagents gave exclusively the α-epoxide **26.15** (Fig. 1), whereas epoxidation of the hindered β-alcohol **26.9** proved to be more intriguing. Thus, treatment of **26.9** with m-CPBA in CH_2Cl_2 yielded a 1 : 1 mixture of two isomeric epoxides **26.16** and **26.17**. In the seemingly less polar solvents hexane, C_6H_6 and Et_2O, the ratio of **26.16** : **26.17** was 9 : 1. By contrast, $VO(acac)_2$/TBHP epoxidation of **26.9** in C_6H_6, under several sets of conditions of temperature and catalyst concentration consistently provided a 60 : 35 : 5 mixture of **26.17**, **26.8**, and the α-OH, α-epoxide **26.15**. In CH_2Cl_2 solution the amount of isomerized epoxy alcohol **26.15** increased relative to the expected ester **26.17** with increasing concentration of the catalyst, while the amount of enone **26.8** remained constant at 15% (see Fig. 1).

The dependence of the ratio **26.17** : **26.15** upon catalyst concentration was explained by invoking displacement of the sterically hindered vanadium complex of axial alcohol **26.9** with catalyst from the α-face, followed by epoxidation from the bottom side of the molecule.

Marasmane Sesquiterpenes

One dozen years after their synthesis of (±)-velleral (**10.6**) WICKBERG's group in Lund (Sweden) brilliantly completed the first enantioselective total synthesis of (+)-isovelleral (**6.2**). The absolute

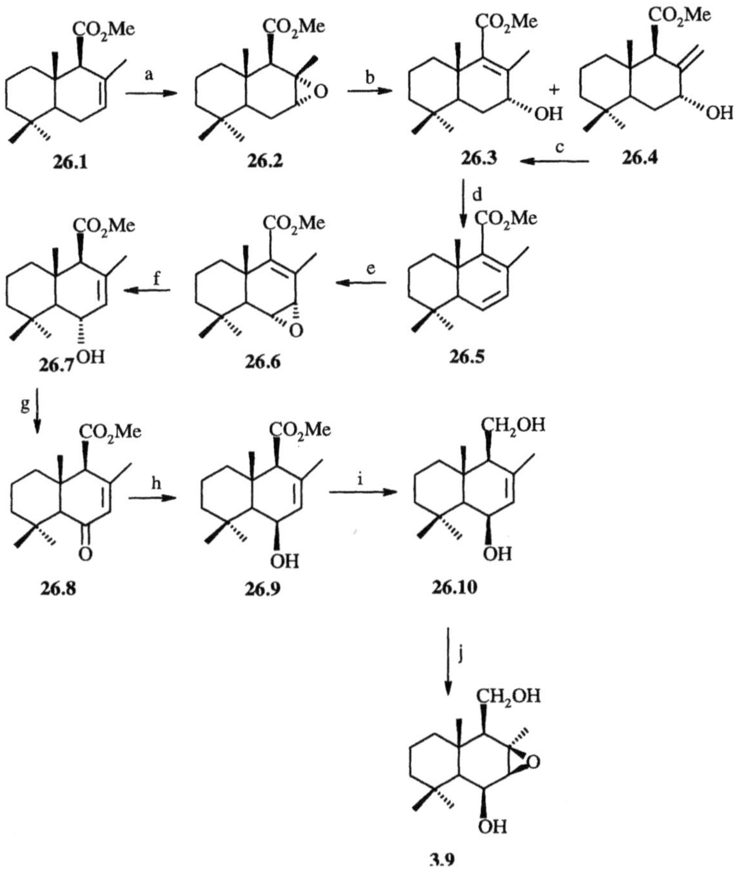

Scheme 24. a) mCPBA, CH_2Cl_2, 78%; b) LDA, THF, −78° −> 25°C, 79%; **26.3 : 26.4**, 4 : 1; c) KOH, dioxane, reflux; d) Ms_2O, Et_3N, DMAP, CH_2Cl_2, reflux, 84%; e) mCPBA, CH_2Cl_2, 25°C, 81%; f) Li/NH_3, THF, −78°C, 52%; g) PCC, CH_2Cl_2, 25°C; h) DIBAL, THF / hexane, −5°C, 91% from **26.7**; i) $LiAlH_4$, Et_2O; j) mCPBA or $VO(acac)_2$ / TBHP, CH_2Cl_2

configurations of this important marasmane sesquiterpene and of others stereochemically correlated with isovelleral were thus definitely confirmed. The synthesis is shown in Scheme 26 and features as a key step a remarkable diastereoselective intramolecular Diels-Alder cyclisation of the chiral cyclopropenyl complex **26.25** derived from D-

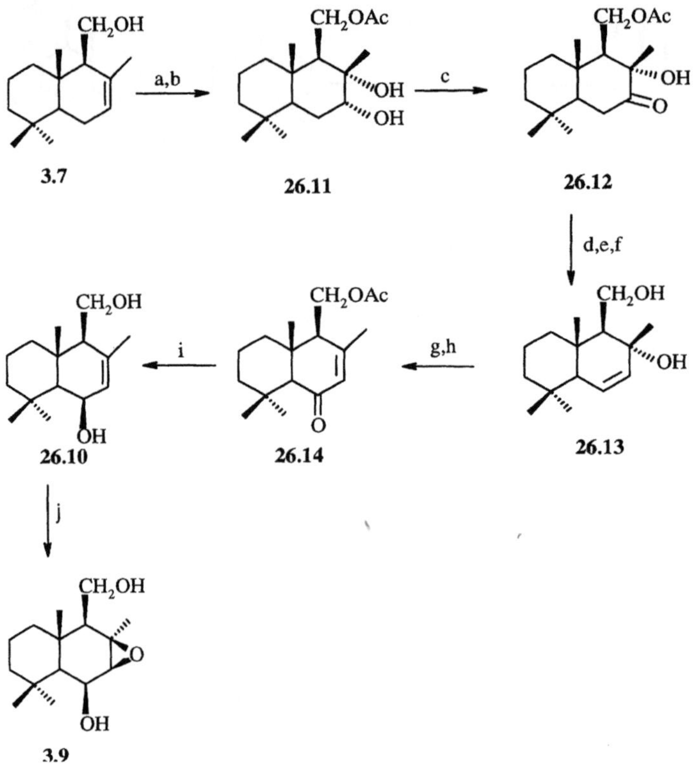

Scheme 25. a) Ac_2O, Py; b) OsO_4, 56% overall yield; c) NBS, CH_2Cl_2, 70%; d) KOH, MeOH; e) $TsNHNH_2$, $BF_3{\cdot}OEt_2$, C_6H_6; f) BuLi, THF, 0°C, 79% for the three steps; g) Ac_2O; h) PCC, 69%; i) DIBAL, THF, 0°C, 98%; j) mCPBA, CH_2Cl_2, 25°C, 95%

ribonolactone (*57*). This cycloaddition occurred exclusively in the *exo*-fashion with the cyclopropenyl group approaching the diene from the α-face, thus affording a cycloadduct (**26.26**) having a stereostructure diastereomeric of isovelleral (**6.2**). However, in a later step (**6.10**↔**6.2**), a reversible thermal rearrangement reaction allowed to invert the stereorelationship of the cyclopropane ring with respect to the bridgehead hydrogens and afforded (+)-isovelleral in good yield after recycling the thermodynamically favored isoisovelleral (**6.10**). The mechanism of this electrocyclic isomerisation has already been discussed in Part 25.

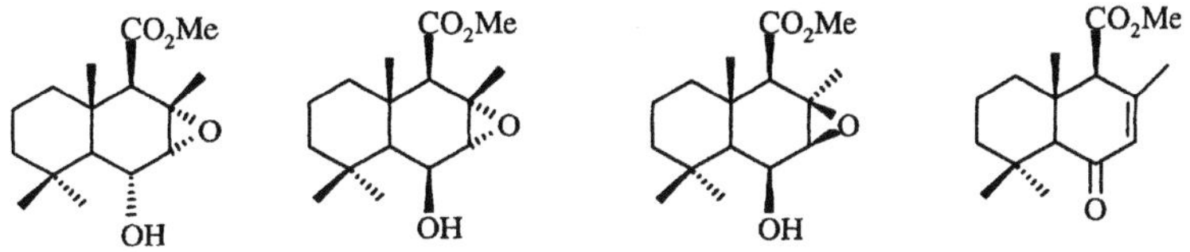

26.15 **26.16** **26.17** **26.8**

Fig. 1

The total synthesis of (+)-isovelleral (**6.2**) also constituted a formal synthesis of (+)-velutinal (**7.12**)) (*72*).

A second highly stereoselective and efficient synthesis of isovelleral, though in racemic form, was published by HEATHCOCK and THOMPSON in preliminary form in 1990 (*150*) and as a full paper in 1992 (*145*), along with an efficient synthetic route to (±)stearoylvelutinal (**7.30**). The synthesis of (±) **6.2** is summarised in Scheme 27.

The synthesis of racemic stearoylvelutinal (**7.30**) diverged from the enol triflate **26.40**. A variant of the Stille coupling afforded a compound (**26.42**) possessing two carbonyls of different reactivity and therefore easily transformed into lactone **7.3**. This was converted into **7.30** in three additional steps (Scheme 28).

A noteworthy feature of the last two syntheses is that they do not require the use of any protecting group. Moreover, they can be readily adapted through the use of ^{13}CO or ^{14}CO in the Stille coupling to provide labelled material to be used in biosynthetic studies.

During these studies, HEATHCOCK made the interesting observation that, contrary to an earlier statement (*36*), synthetic velutinal (**7.12**), as well as a sample prepared by basic ethanolysis (EtO^-/ EtOH) of natural esters **7.28** or **7.30** of velutinal (*37*) existed as a 5.5 : 1 mixture of β : α isomers at the anomeric centre (as determined by NMR spectral data). In addition, the O-methyl acetal of velutinal (**7.17**), prepared by neutral methanolysis of **7.30**, also existed as an inseparable 4 : 1 mixture of anomers, the β-diastereomer again prevailing. In contrast, natural stearoylvelutinal (**7.30**) exists as only one isomer and the relative configuration of the acyloxy group was assigned as *cis* to the cyclopropane ring (NOESY spectrum). The same diastereomer, uncontaminated by the other anomer, was obtained by acylation of (±)-velutinal with stearoyl chloride under basic conditions (see Scheme 28). It has been demonstrated that this stereochemistry resulted from a more rapid acylation of the β rather than the α-anomer of free velutinal (**7.12**) with concomitant fast equilibration of the two isomers.

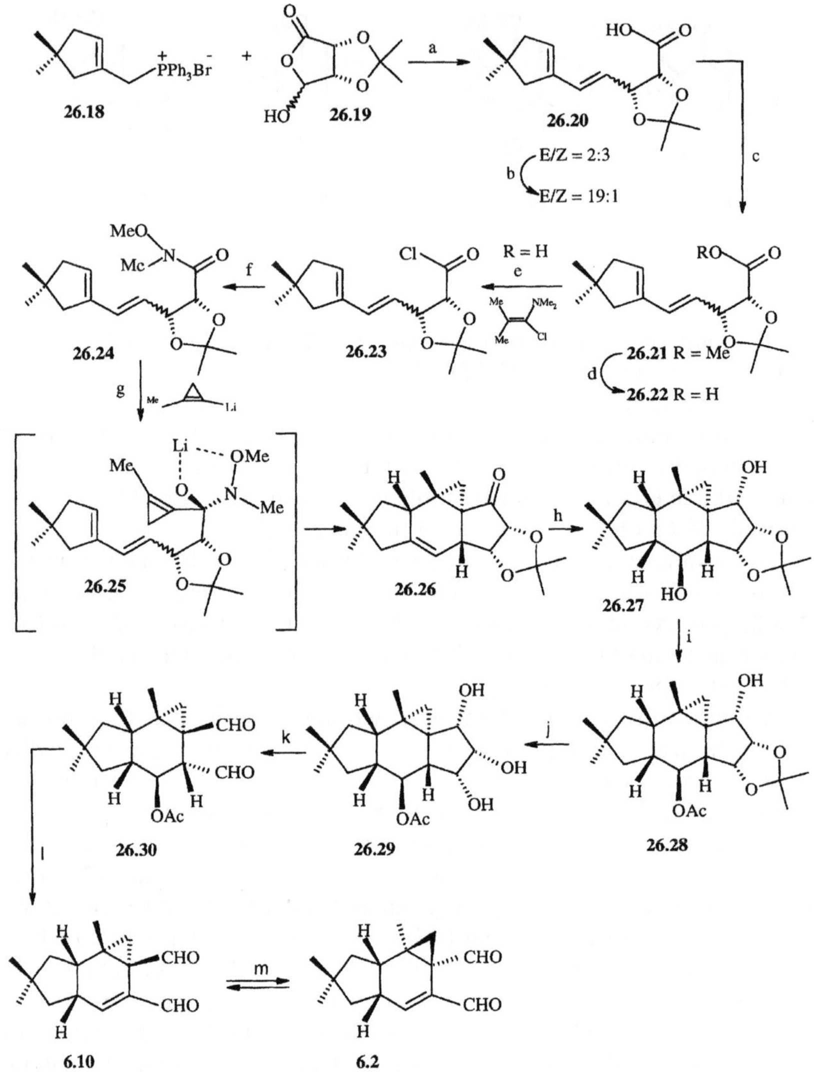

Scheme 26. a) **26.18**, THF, BuLi, 2°C, then −70°C, **26.19**, 15 min, reflux, 2 h; b) $Hg(OAc)_2$, MeOH, 22°C, 60 H, then Zn dust, 10 min; c) CH_2N_2, Et_2O, then column chromatography, 60% from **26.18**; d) NaOH, MeOH-H_2O, 97%; e) CH_2Cl_2, 2 min, 22°C; f) MeONHMe·HCl, 2°C, Et_3N, → 22°C, 0.5 h, 85%; g) Et_2O, −70°C, then → 22°C, 68%; h) BH_3·THF, THF, 22°C, 6 h, then NaOH in H_2O, H_2O_2, 22°C, 0.5 h, 77%; i) Py, DMAP, CH_2Cl_2, Ac_2O, −70 → 22°C, 1 h, 86%; j) H_2SO_4 (0.2 M) in MeOH-H_2O (4 : 1), 22°C, 20 h, 75%, and **26.28**, 8%; k) $NaIO_4$, EtOH, 22°C, 0.5 h; l) Py, reflux, 0.5 h, 55% from **26.29**; m) mesitylene, reflux, 0.5 h (**6.10 : 6.2**, ca 3 : 2), column chromatography, then **6.10** is recycled, 71% of **6.2** after five cycles

References, pp. 161–171

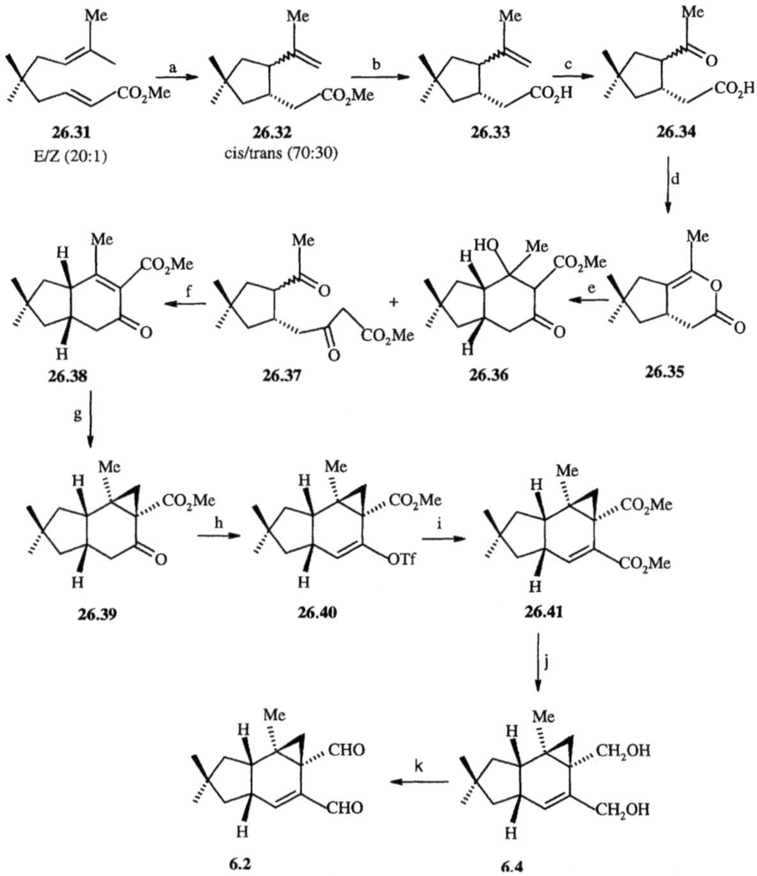

Scheme 27. a) Neat, Ar atmosphere, Pyrex bomb, 235°C, 24 h, 93%; b) 3 M aq KOH, MeOH, 22 h, 100%; c) O_3, $MeOH/CH_2Cl_2$ (1 : 1), −78°C, then Me_2S, −78 → 25°C, 100%; d) $(COCl)_2$, C_6H_6, reflux, 22 h, 94%; e) $LiCH_2CO_2Me$, THF, −78°C, 1 h; f) $MeSO_3H$, C_6H_6, 25°C, 5 min, 50% from **26.35** g) $Me_2S(O)CH_2$, THF, 25°C, 30 min, 65%; h LDA, THF, −78°C, 40 min, then $PhNTf_2$ in THF, −78 → 25°C, 1 h, 98%; i) $Pd(OAc)_2$, PPh_3Et_3N, MeOH, DMF, CO atmosphere, 25°C, 2 h, 93%; j) DIBAL, toluene-THF, −78 → 25°C, 30 min, 100%; k) $(COCl)_2$, CH_2Cl_2, DMSO, −78°C, then NEt_3, −78 → 25°C, 83%

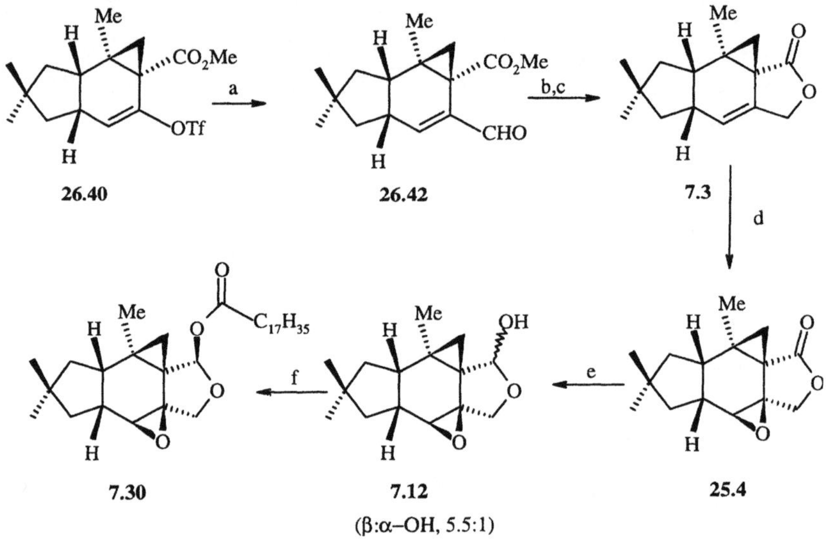

Scheme 28. a) Cat $Pd(PPh_3)_4$, CO (3 atm), Bu_3SnH, excess LiCl, THF, 50°C, 88%; b) $NaBH_4$, $CeCl_3 \cdot 7\ H_2O$, EtOH; c) *p*-TsOH, C_6H_6, 45°C, 10 min, 93% from **26.42**; d) dimethyldioxirane, Me_2CO, 0°C, 84%; e) 1.3 eq DIBAL, PhMe, −78°C, 3 h, 50% isolated yield; f) n-$C_{17}H_{35}COCl$, Et_3N, CH_2Cl_2, 0°C, 5 min, 63% isolated yield

Lactarane and Lactaranolide Sesquiterpenes

Pioneering synthetic work by G. MAGNUSSON and collaborators on lactarane sesquiterpenes culminated in the first syntheses of racemic velleral (**10.6**), vellerolactone (**11.6**), pyrovellerolactone (**11.7**), and stereoisomers and regioisomers of these. These efforts were of seminal importance because they led to revision of the previously assumed structures of these *Lactarius* sesquiterpenes and their synthetic isomers. In fact, the Swedish authors unequivocally demonstrated that the Me-12 group in velleral (**10.6**) and vellerolactone (**11.6**) is situated *cis* to the ring junction hydrogen atoms and not in the *trans* arrangement previously reported, and that the carbonyl group in **11.6** and **11.7** is situated on C-5 instead of C-13. For the sake of clarity, in Fig. 2 and Scheme 29 we show the correct structures of these natural products, their isomers and related compounds that were synthesized by the Swedish group, with references to the original literature.

To summarize, a general synthetic route was worked out and successfully carried through by G. MAGNUSSON to prepare natural and synthetic vellerane derivates with *cis* hydrogens at the ring junction and a

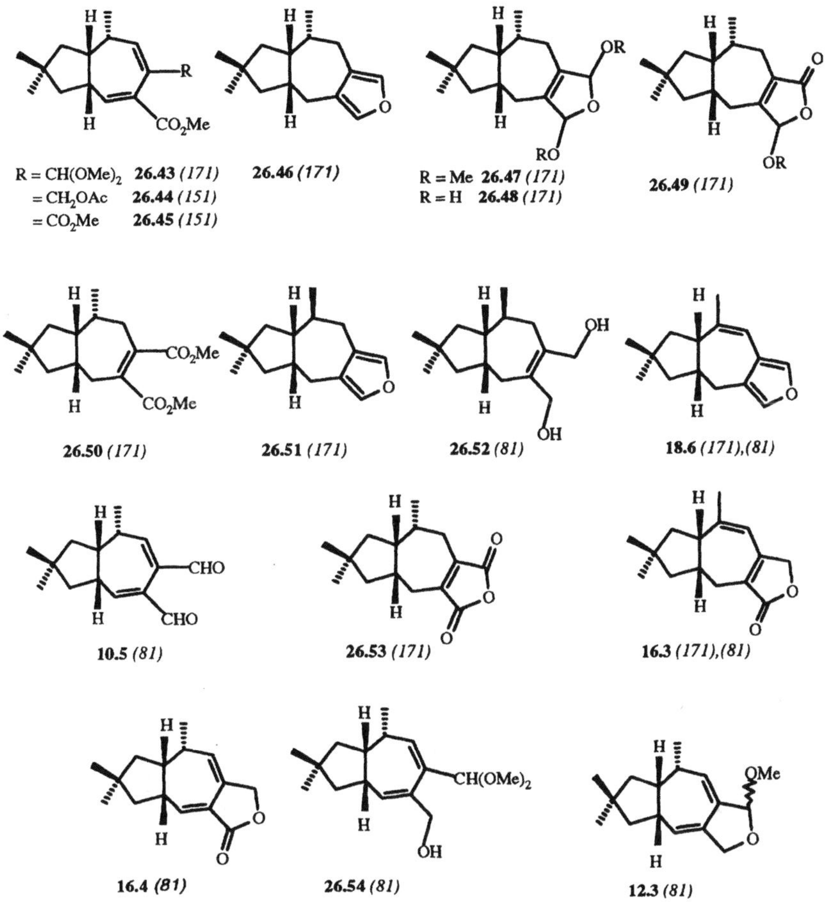

Fig. 2

C-3 methyl group either *cis* or *trans* to these, and the lactone carbonyl group (where applicable) either at C-5 or at C-13. Key steps for assembling the lactarane ring system was a [2+2] cycloaddition reaction between the pyrrolidine enamine of a methyl-substituted *cis*-bicyclo [3.3.0] octanone (*81*, *151*) and an appropriately 4-substituted tetrolic ester, followed by fission of the thus generated four membered ring and subsequent hydrogenolytic deamination with diborane. Further functional group manipulation gave the target compound. Syntheses of racemic velleral (**10.6**), vellerolactone (**11.6**), and pyrovellerolactone (**11.7**) exemplify this strategy (Scheme 29) (*81*).

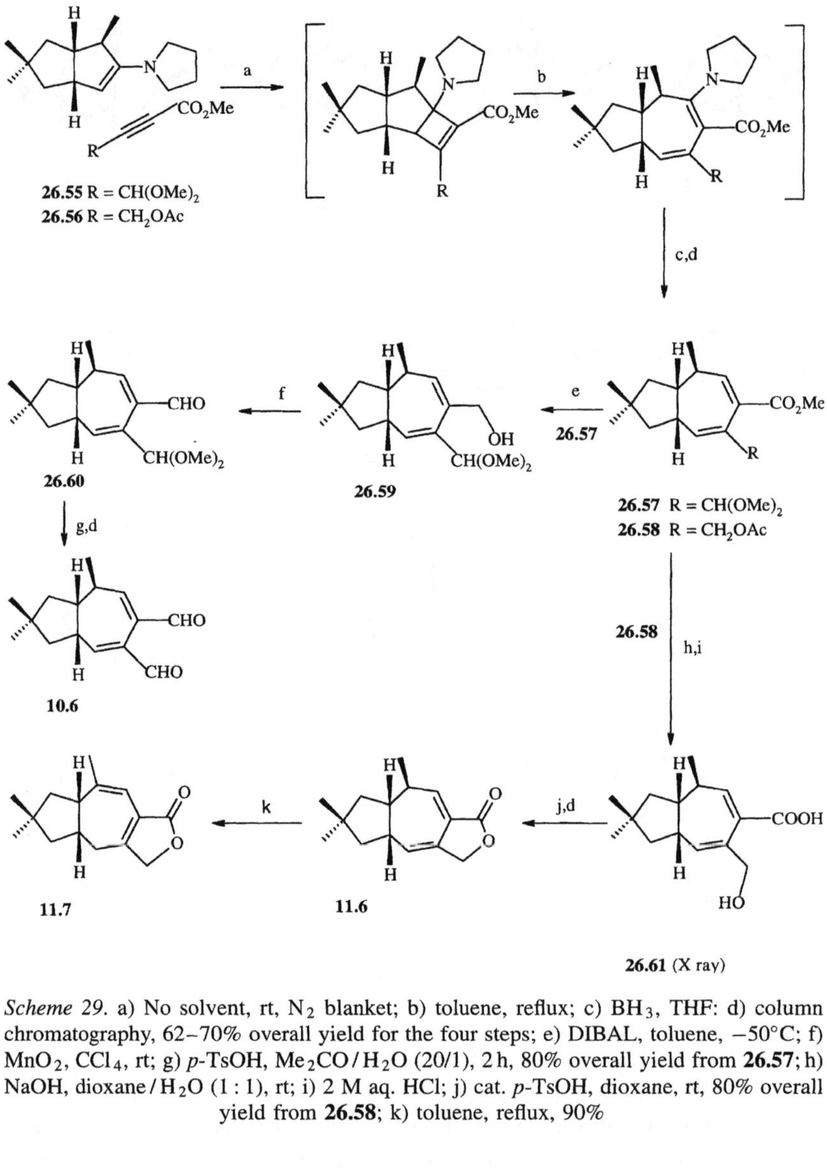

Scheme 29. a) No solvent, rt, N_2 blanket; b) toluene, reflux; c) BH_3, THF: d) column chromatography, 62–70% overall yield for the four steps; e) DIBAL, toluene, −50°C; f) MnO_2, CCl_4, rt; g) *p*-TsOH, Me_2CO/H_2O (20/1), 2 h, 80% overall yield from **26.57**; h) NaOH, dioxane/H_2O (1 : 1), rt; i) 2 M aq. HCl; j) cat. *p*-TsOH, dioxane, rt, 80% overall yield from **26.58**; k) toluene, reflux, 90%

After preliminary communications (*152, 153*) on a new approach to the lactarane skeleton based on the Diels-Alder reaction of a furan derivative with maleic anhydride, TOCHTERMANN in 1997 reported the synthesis of 2(3)-8(9)-bisanhydrolactarorufin A (**11.3**) (*154*) (Scheme

30). Cycloaddition reaction of furan **26.62** with dimethyl acetylenedicarboxylate (**26.63**) followed by selective oxidation of the tetralkyl substituted double bond of the resulting cycloadduct **26.64** gave diketodiester **26.65**, which underwent a highly regioselective aldol condensation to afford hydroazulenone **26.66**. This compound gave tropone **26.67** on treatment with Me_2BBr/Et_3N. It is worth noting that all attempts to cleave the ether bridge between C-4 and C-8 of such hydroazulenones with a C(6)–C(7) single bond failed.

Preferential catalytic hydrogenation of the C(4)–C(6) double bond of **26.67**, followed by methylenation with the Nysted reagent (Aldrichimica Acta **26**, 14 (1993), completed the synthesis of lactarane diester **26.69** which afforded the crystalline target compound **11.3** in two additional steps.

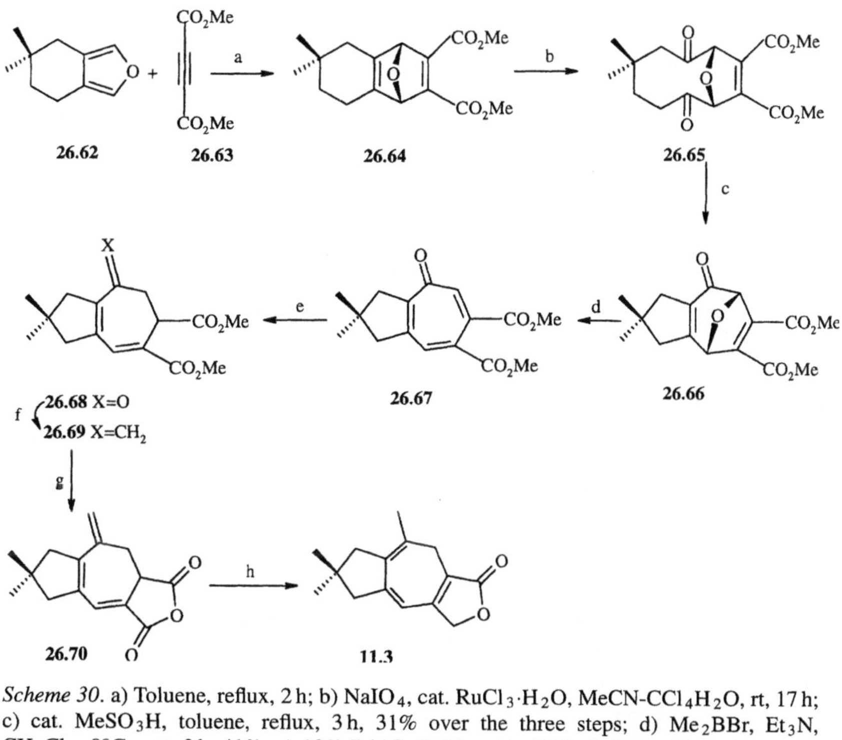

Scheme 30. a) Toluene, reflux, 2 h; b) $NaIO_4$, cat. $RuCl_3 \cdot H_2O$, MeCN-CCl_4H_2O, rt, 17 h; c) cat. $MeSO_3H$, toluene, reflux, 3 h, 31% over the three steps; d) Me_2BBr, Et_3N, CH_2Cl_2, 0°C → rt, 3 h, 41%; e) 10% Pd/C, EtOAc, 4 h, 53%; f) $TiCl_4$, Nysted reagent, THF, −78°C → rt over 3 h, 59%; g) i) NaOH, MeOH-H_2O, reflux, 2 h; ii) Ac_2O, 80°C, 0.5 h; h) i) $NaBH_4$, THF, 0°C, 30 min, then 60 min at rt; ii) 2N HCl, 24% over the steps g) and h)

Furanolactarane Sesquiterpenes

Two syntheses of racemic furanether B (**18.8**) were described by N.E. SCHORE *et al.* (*155*), (*156*). Both feature an octacarbonyldicobalt catalysed cycloaddition of 8-oxabicyclo{3.2.1}oct-6-ene derivatives with alkynes (Pauson-Khand reaction) as a key step. The bicyclic ketone **26.71**, prepared by [4+3] cycloannulation from 2-methyfuran and tetrabromoacetone, was used as starting material.

In the first approach (Scheme 31) (*155*), compound **26.71** was reduced first to differentiate the carbonyl initially present from one produced later in the synthesis. Pauson-Khand reaction of the mixture of stereoisomeric alcohols **26.72** gave a mixture of four isomeric *anti* and *syn* ketones, **26.73a**, **26.74a** and **26.73b**, **26.74b**, respectively. Notwithstanding the small degree of regioselectivity observed (*anti* : *syn* *ca.* 1.5 : 1) the reaction was completely stereoselective, leading to compounds possessing the exo configuration at the newly formed ring fusion (furan ether β stereochemistry). To reduce synthetic complications the mixture was separated into exo (**26.73a** +**26.73b**) and endo (**26.74a** +**26.74b**) pairs of *anti* and *syn* regioisomers, and each pair was carried through as a mixture, to give eventually the same synthetic intermediate **26.76**. Details concerning the further synthetic transformations of cycloadducts **26.73a,b** and **26.74a,b** are presented only for the two exo-alcohols in Scheme 32, since the endo isomers were also converted into ketone **26.76** by a similar synthetic sequence, albeit in lower overall yield.

Of the three isomeric alcohols **26.75a**, **26.75b**, and **26.75c** in Scheme 32 only the first one was then taken further in the synthesis, although in principle the other two compounds could also be converted into ketone **26.76** by analogous reactions. Annulation of the furan ring onto bicyclic ketone **26.76** was accomplished by initial regiospecific formylation to give **26.77**, followed by an original reaction sequence worked out by the authors (*157*), which provided furanether B (**18.8**) in reasonable overall yield.

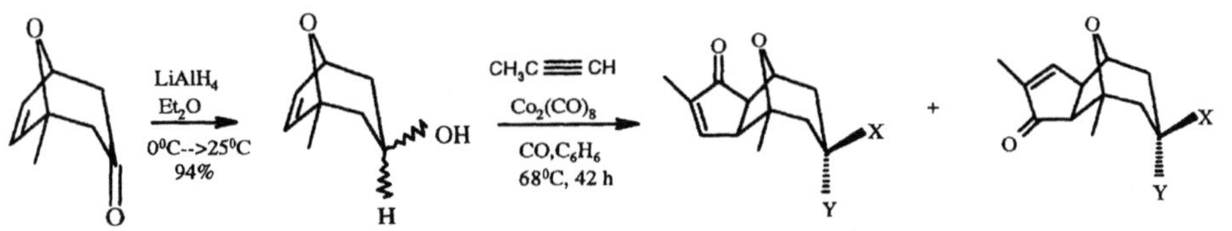

26.71 **26.72** (exo:endo, 2:1) *anti* *syn*

26.73a X=OH; Y=H **26.73b** X=OH; Y=H
26.74a X=H; Y=OH **26.74b** X=H; Y=OH

Scheme 31

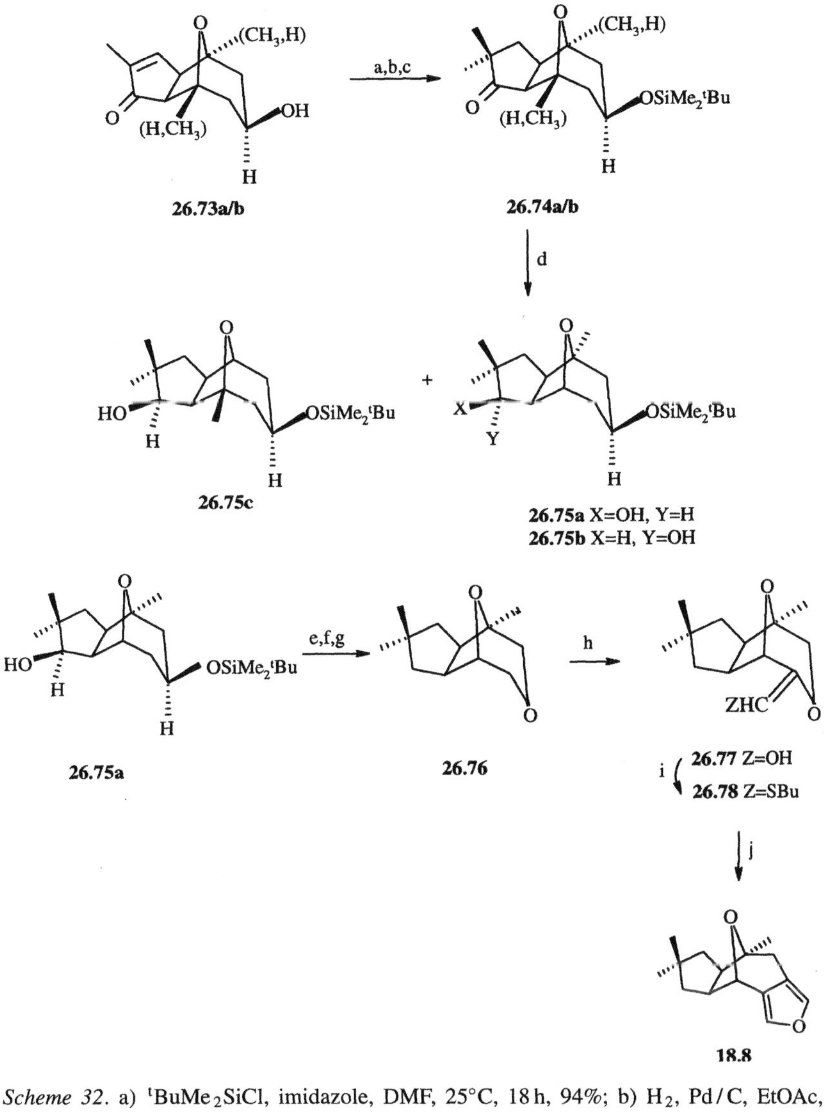

Scheme 32. a) $^{t}BuMe_2SiCl$, imidazole, DMF, 25°C, 18 h, 94%; b) H_2, Pd/C, EtOAc, MeOH, 25°C, 10 h, 100%; c) MeI, $KO^{t}Bu$, $^{t}BuOH$, C_6H_6, 25°C, 20 min → 38°C, 35 min, 100%; d) $LiAlH_4$, Et_2O, 25°C, 1.5 h, 100%; e) NaH, THF, 66°C, 1 h, then NaH, CS_2, THF, 66°C, 0.5 h, then MeI, 66°C, 0.5 h, 100%; f) Bu_3SnH, $C_6H_5CH_3$, 111°C, 16 h, then Bu_3SnH, AIBN, $C_6H_5CH_3$, 111°C, 6 h, 50%; g) PCC, CH_2Cl_2, 25°C, 70 min, 60%; h) HCO_2Et, NaOMe, C_6H_6, 25°C, 18 h, 67%; i) BuSH, *p*-TsOH, C_6H_6, 80°C, 1 h, 100%; j) $Me_3S^+MeSO_4^-$, CH_2Cl_2, 50% aq. NaOH, 48°C, 24 h, then concentration, 25°C, 24 h, then aq HCl, THF, 25°C, 3 h, 70%

In a subsequent approach (*156*) SCHORE found that the simple expedient of constructing the furan moiety early in the reaction sequence eliminated the problems associated with differentiating the carbonyl group in **26.71** from that introduced later in the Pauson-Khand cycloaddition reaction. A much shorter synthetic pathway was carried out which completely avoided the use of protecting groups or other excessive functional group manipulation (Scheme 33).

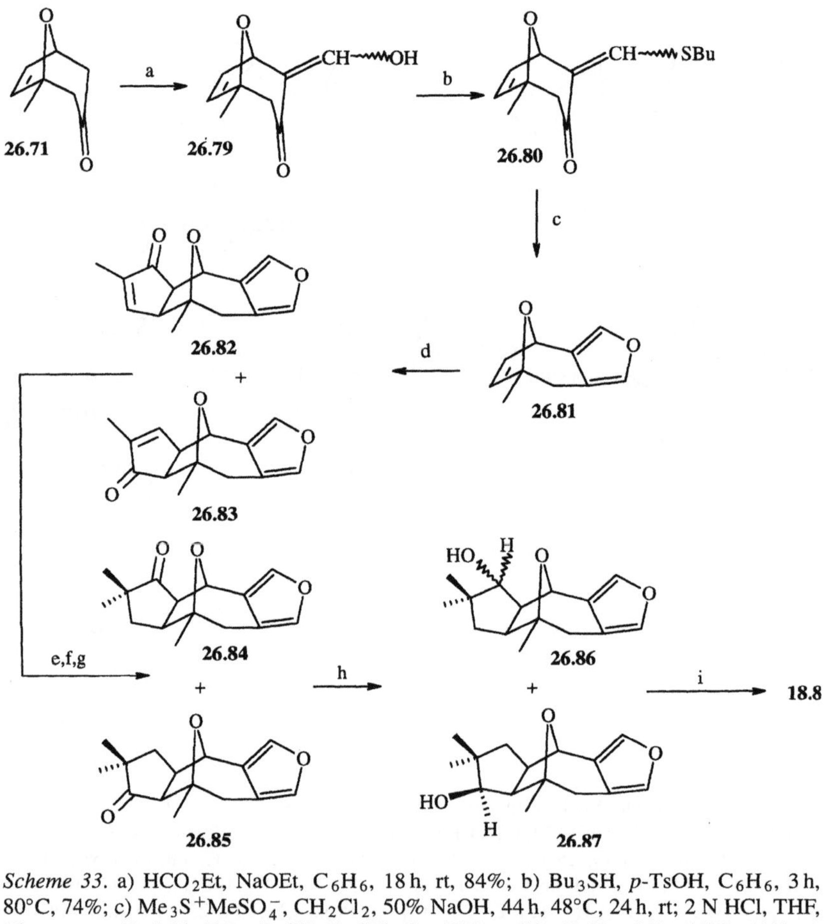

Scheme 33. a) HCO_2Et, NaOEt, C_6H_6, 18 h, rt, 84%; b) Bu_3SH, *p*-TsOH, C_6H_6, 3 h, 80°C, 74%; c) $Me_3S^+MeSO_4^-$, CH_2Cl_2, 50% NaOH, 44 h, 48°C, 24 h, rt; 2 N HCl, THF, 3 h, rt, overall yield 45%; d) $CH_3C{\equiv}CH$, $Co_2(CO)_8$, CO, C_6H_6, 44 h, 64°C, 64%; e) $LiAl(O^tBu)_3H$, THF, 30 h, 5°C; f) PCC, CH_2Cl_2, 45 min, 62%; g) CH_3I, 2 eq KO^tBu, ^tBuOH, C_6H_6, 45%; h) $LiAlH_4$, Et_2O, 30 min, 0°C, 1 h, rt, 75%; i) NaH, CS_2, CH_3I, Bu_3SnH, $C_6H_5CH_3$, 30 min, 110°C, 71%

Pauson-Khand annulation of furan **26.81** provided the two *exo* regioisomers **26.82** and **26.83** in a 2 : 1 ratio. The entire mixture was then carried through the following steps since both isomers resulted in the same final product, furanether B (**18.8**). Noteworthy in this approach was the demonstration of tolerance of the furan ring to organometallic cycloaddition conditions and the selective enone reduction in a five-membered ring (steps d and e, Scheme 33).

For the synthesis of the 8-oxabicyclo [3.2.1] octane ring system present in furanether B (**18.8**) G.A. MOLANDER and collaborators (*158*) relied upon the Lewis acid-promoted [3+4] annulation of 1,4-dicarbonyl compound **26.93** with bisnucleophilic bis (trimethylsilyl)enolether **26.94**. This reaction efficiently produced the tricyclic ether **26.95** possessing the desired stereochemistry of the natural product (**18.8**), as well as having the β-ketone-ester functionality suitably placed for the construction of the furan ring of furanether B (Scheme 34). To complete the synthesis, compound **26.95** was readily converted to the enol triflate **26.96**, which underwent a Pd catalysed carbonylation to afford the butenolide **26.97** in high yield. Reduction of **26.97** with DIBAL in CH_2Cl_2 provided lactol **26.98** which was not isolated but directly converted into **18.8**. On the other hand, decarboxylation of keto ester **26.95** provided tricyclic ketone **26.76**, identical with that synthesised by SCHORE using a completely different strategy (see Scheme 32).

A fourth approach to the oxatricyclo-[5.3.1.$0^{2,6}$]-undecane ring system of furanether B (**18.8**) was published recently by de Groot *et al.* (*159*) who relied upon a stereoselective base-induced rearrangement reaction of 1,4-diol monosulfonate esters to establish the bridged ether core of **18.8** (**26.103**→**26.105**, Scheme 35). The starting material of the synthesis was the known ketone **26.99** which was converted to the required methanesulfonate **26.103** by standard procedures.

Upon exposure to $LiAl(O^tBu)_3H$ (2.5 equiv) in refluxing toluene, ketone **26.103** smoothly rearranged to a 10 : 1 mixture of **26.105** and **26.106** *via* the intermediate alkoxide **26.104**. The tricyclic ether **26.105** was then converted to SCHORE's ketone **26.76**, thereby completing a formal total synthesis of racemic furanether B (**18.8**). In addition, DE GROOT *et al.* explored an alternative conversion of **26.76** to **18.8** *via* the isomeric lactaranolides **26.97** and **26.110**, which could be obtained in fair yield using the Pummerer-induced cyclization reaction of sulfoxide **26.108** as a key reaction (Scheme 36).

The first total synthesis of (±)-furoscrobiculin B (**18.9**) was accomplished by K. KANEMATSU (*160*) through a novel construction of the azuleno [6,7-*c*] furan ring system by base catalysed pinacol-type rearrangement of the isonaphthofuran derivatives **26.122** (Schemes 37

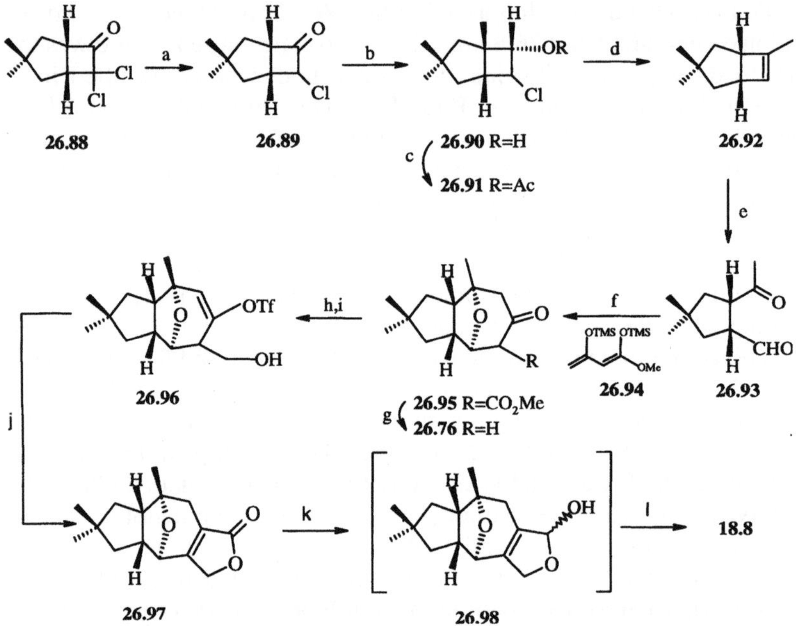

Scheme 34. a) Zn, AcOH, rt, 94%; b) MeMgBr, Et_2O, −78°C, 87%; c) Ac_2O, NEt_3, DMAP, CH_2Cl_2, rt, 85%; d) Na, NH_3, Et_2O, −78°C; e) O_3, pentane-EtOAc, cat. $NaHCO_3$, −78°C, then PPh_3, −78°C to rt, 47% from **26.91**; f) cat. TMSOTf, CH_2Cl_2, −78°C, 70%; g) NaCl, H_2O, DMSO, 140°C, 65%; h) NaH, 2-N(SO_2CF_3)$_2$-5-C–C_5H_3N, THF, 0°C to rt, 78%; i) DIBAL, THF, −55°C to rt, 85%; j) cat. Pd(PPh_3)$_4$, CO, NBu_3, LiCl, CH_3CN, 60°C, 90%; k) DIBAL, CH_2Cl_2, −78°C, then l) 1N H_2SO_4, −78°C to rt, 90% for the two steps

and 38). This compound was prepared in a convergent way from stereoisomeric tricyclic furans **26.118** and **26.119** arising from the Furan Ring Transfer (FRT) reaction on hydrobenzofuran **26.115**.

Following this first approach, an improved synthetic route to (±)**18.9** was later developed by KANEMATSU and collaborators (Scheme 39) (*161*). Prop-2-ynyl-ether **26.132**, obtained in a few steps from the known 1,2-diketone **26.127**, was submitted to the FRT reaction to yield naphto-[2,3-*c*]furandiol **26.133** in good yield. The latter compound was then converted to the stereochemically desidered cis-diol **26.137**, via ketone **26.136** which underwent diastereoselective reduction from the α-side when using $Zn(BH_4)_2$ at −100°C.

Pinacol-type rearrangement of *cis*-diol **26.137** by esterification with TsCl in pyridine at rt readily proceeded with high stereoselectivity to afford directly the azulenofuran **26.138** (Scheme 40). However, direct

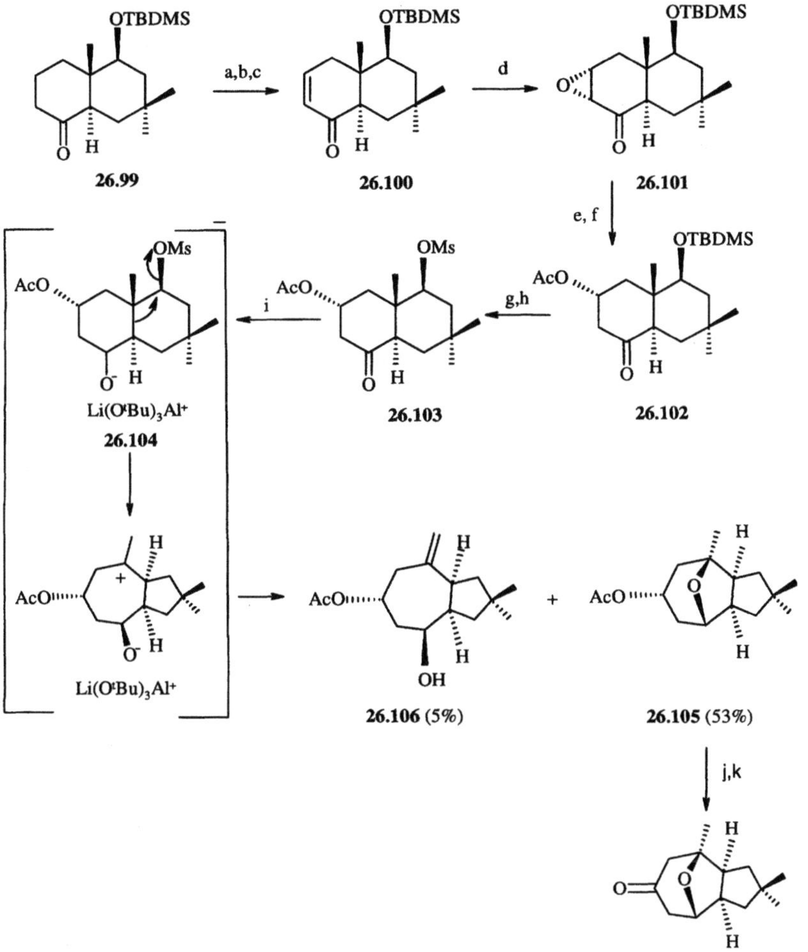

Scheme 35. a) LDA, THF, −78°C, then TMSCl; b) NBS, THF, 0°C to rt; c) Li_2CO_3, LiBr, DMF, 140°C, 94% for the three steps; d) 35% H_2O_2, 1 M aq NaOH, MeOH, 71%; e) 3 equiv Me_2CuLi, Et_2O, −25°C, 64%; f) Ac_2O, Py, DMAP, 94%; g) 40% HF, MeCN, rt, 95%; h) MsCl, Py, rt, 99%; i) 2.5 equiv. $LiAl(O^tBu)_3H$, $C_6H_5CH_3$, reflux; j) $LiAlH_4$, Et_2O, rt, 81%; k) PDC, CH_2Cl_2, rt, 93%

methylation of compound **26.138** using MeLi gave the undesired 3-*epi*-furoscrobiculin B **26.126** as the major product, which might be caused by attack of the nucleophile from the convex face of the substrate. Therefore, βγ-enone **26.138** was at first isomerized to the corresponding

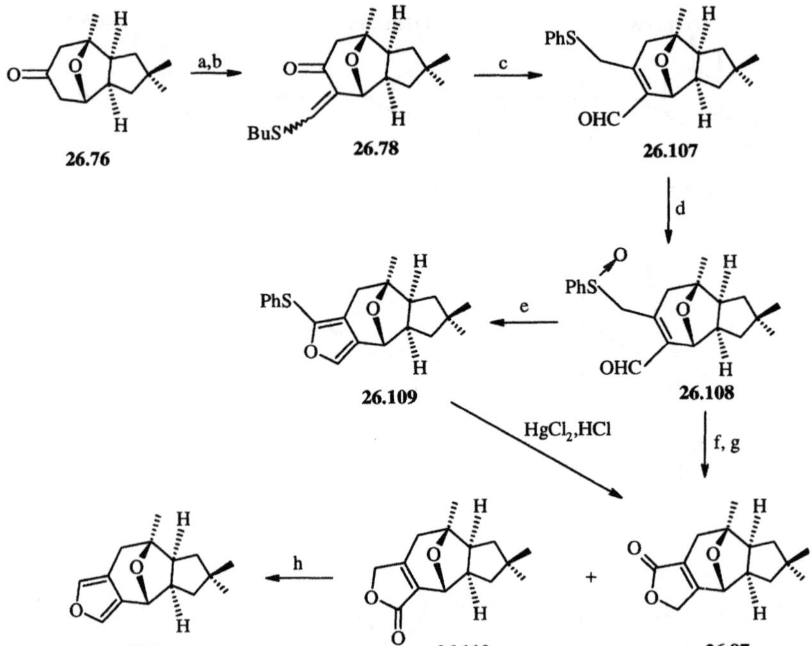

Scheme 36. a) NaH, Et_2O-MeOH, HCO_2Et, rt, 95%; b) BuSH, *p*-TsOH, C_6H_6, 88%; c) $PhSCH_2CeCl_2$, THF-Et_2O, −78°C, then $HgCl_2$, EtOH-H_2O, rt, 84% overall yield; d) $NaIO_4$, MeOH-H_2O, rt, 3 d, 99%; e) 1 equiv. 2,6-lutidine, 1 equiv. TFAA, CH_2Cl_2, rt, 25 min, 43%; f) excess 2,6-lutidine, excess TFAA, CH_2Cl_2, −25°C, 44 h; g) $HgCl_2$, 4 M aq HCl, 35°C, 80% from **26.108**, ca. 4 : 1 mixture of **26.97** and **26.110**; h) see steps i) and j) in Scheme 34

αβ-enone **26.124** which was then converted to racemic furoscrobiculin B (**18.9**) according to a standard procedure.

8,9-Secofuranolactarane Sesquiterpenes

Two similar approaches (Schemes 41 and 42) were described for the synthesis of racemic lactaral (**19.1**). Both converged to the preparation of the THP ether of lactarol (**19.4**), which was then converted easily into lactaral using standard methods. In the first of these syntheses (Scheme 41) (*128*) the mesitoate **26.139**, prepared from diethyl furan-3,4-dicarboxylate by conventional steps, was coupled with the lithium derivative of the allylic bromide **26.140** to give **19.4**, albeit in very low yield. A much more efficient synthesis of **19.4** (*162*) was completed by coupling chloride **26.142** with the Grignard reagent **26.141** in the

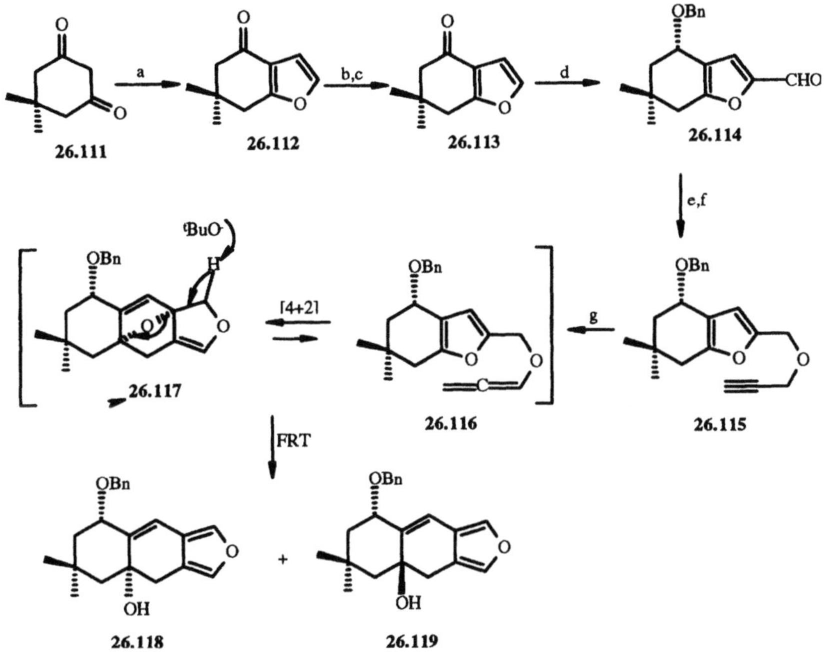

Scheme 37. a) $ClCH_2CHO$, aq. $NaHCO_3$, $CHCl_3$, rt; aq H_2SO_4, rt, 89%; b) $NaBH_4$, EtOH, rt, 94%; c) BnBr, NaH, DMF, rt, 99%; d) $POCl_3$, DMF, rt, 95%; e) $NaBH_4$, EtOH, rt, 100%; f) $BrCH_2C{\equiv}CH$, cat. Bu_4NHSO_4, aq NaOH, Et_2O, rt, 98%; g) ^tBuOK, ^tBuOH, 70°C, 87% (**26.118** : **26.119** = 1 : 3)

presence of Li_2CuCl_4 (Scheme 42). The above syntheses of lactaral (**19.1**), furanether B (**18.8**), and furoscrobiculin B (**18.9**) definitely confirmed the structures of these natural furanosesquiterpenes, which had been assigned on the ground of spectra interpretation.

Part 27. Biological Properties of Metabolites of *Lactarius* Origin

Having read the review, which described the constituents of *Lactarius* mushrooms we can realize that the mushrooms possess a very potent biochemical system, which produces a big variety of compounds. Enzymatic system of majority of species is capable of transforming the velutinal esters into a series of biologically active compounds, which can serve a particular species as defense weapons.

As the biological activities of metabolites from *Lactarius* species have been thoroughly reviewed in a recent publication by P. VITA-FINZI

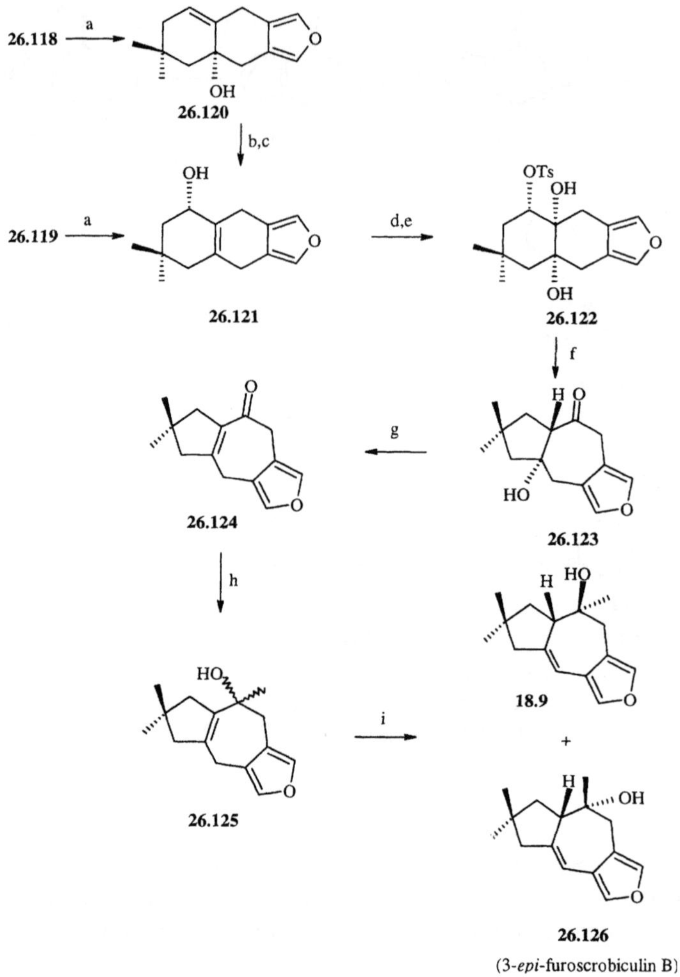

Scheme 38. a) Li, liquid NH_3, THF, −78°C, 80–90%; b) PDC, CH_2Cl_2, rt, 46%; c) DIBAL, THF, −78°C, 85%; d) OsO_4, CH_2Cl_2, Py, rt, 33%; e) TsCl, NEt_3, DMAP, CH_2Cl_2, 40°C, 80%; f) ^tBuOK, ^tBuOH, rt, 97%; g) Al_2O_3, Py, 110°C, 69% (containing 20% β,γ-enone); h) MeLi, $CeCl_3$, THF, −78°C, 79%; i) ^tBuOK, DMF, rt, 68%, **18.9** : **26.126** = 7 : 2)

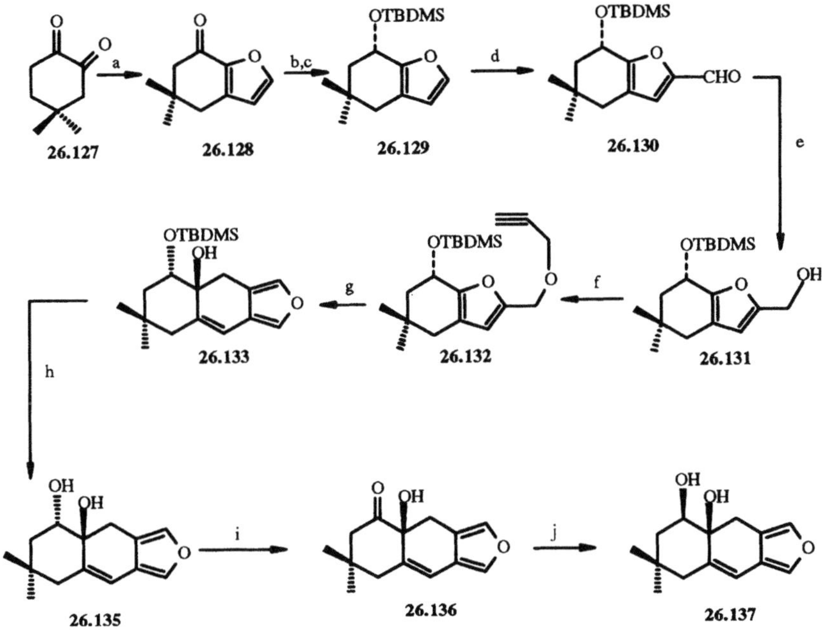

Scheme 39. a) $ClCH_2CHO$, aq. $NaHCO_3$, MeOH, rt; aq. H_2SO_4, rt, 68%; b) $NaBH_4$, EtOH, rt, 93%; c) $^tBuMe_2SiCl$, imidazole, DMF, rt, 92%; d) BuLi, THF, 0°C; DMF, −78°C; e) $NaBH_4$, EtOH, rt, 87% for two steps; f) $BrCH_2C{\equiv}CH$, cat. Bu_4NHSO_4, aq. NaOH, Et_2O, 92%; g) ^tBuOK, ^tBuOH, 80°C, 77% (+11% *cis* isomer); h) TBAF, THF, rt, quant.; i) Dess-Martin periodinane, CH_2Cl_2, rt, 78%; j) 1.5 eq $Zn(BH_4)_2$, Et_2O, −100°C, 98% (dr 93 : 7)

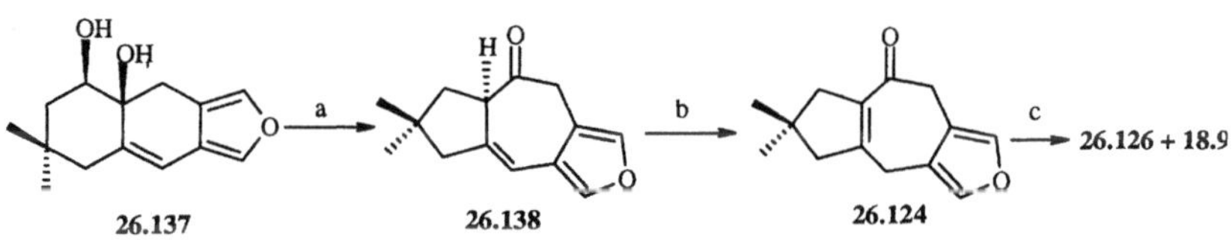

Scheme 40. a) *p*-TsCl, Py, DMAP, rt, 64% (+2% **26.124**); b) NEt_3, Py, rt, 55%; c) see Scheme 38, steps h and i

and G. VIDARI (*16*), only papers which appeared after 1994 will be mentioned at length although references to earlier work will be given. The largest group of papers has dealt with antifeedant activity against the storage pests *Tribolium confusum* Duv., *Trogoderma granarium* Ev. and *Sitophylus granarius* L. (*97, 115, 116, 163, 164*) as well as against mammals (*41*), insects and fishes (*22, 38*). The mutagenic effect of various mushrooms in the Ames *Salmonella typhimurium* tester strains

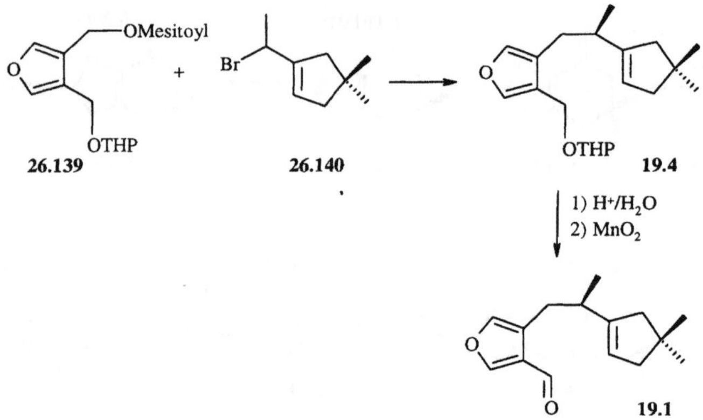

Scheme 41

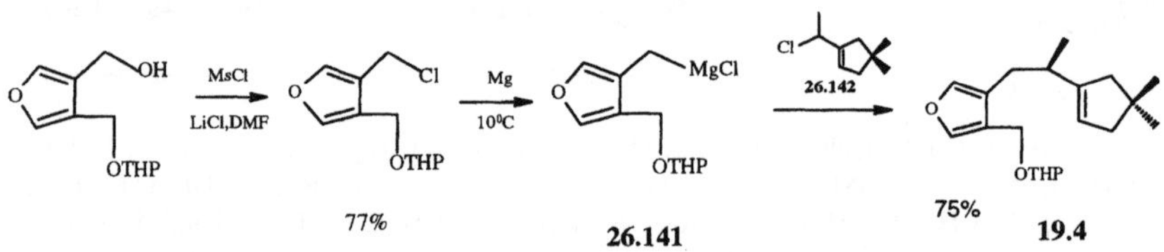

Scheme 42

was studied (*165*, *166*). Cytotoxicity for ECA cells was also reported (*167*). A group of 20 unsaturated dialdehydes isolated from mushrooms and plants showed antibacterial, antifungal, cytotoxic, algaecidal and mutagenic activities (*168*).

Investigations of mutagenic activities of unsaturated dialdehydes have continued. It was found that merulidial, a marasmane sesquiterpene isolated from *Merulius tremellosus* Fr. (*169*) as well as isovelleral (**6.2**) underwent autoxidation, in normal bioassay media to form 9-hydroxy derivatives which were more mutagenic than their parent dialdehydes (*61*, *62*). The rate of autoxidation was much faster in alkaline media. In addition, it was found that natural (+)-isovelleral (**6.2**) was twice as active as the synthetic (racemic) analog, indicating that (−)-isovelleral is inactive or only weakly active. This supports the suggestion that the activity of isovelleral (**6.2**) depends upon the absolute configuration of its cyclopropane dialdehyde moiety and is in agreement with the earlier finding *(61)* that isoisovelleral (**6.10**) had only approximately 10% of the mutagenic activity of **6.2**.

Studies of antifeedant activity against storage pests have continued. A paper reporting antifeedant activities of 53 compounds *(170)* investigated structure-activity relationships. Compounds possessing lactarane and marasmane skeletons were more active than isolactaranes and increasing the number of hydroxy groups present in a molecule decreased its activity. Generally compounds of natural origin, with the exception of keto derivatives, possessed greater activity than their chemically modified analogues. The unusually high antifeedant activity *(170)* of 3-O-ethylfurandiol (**18.26**), an artifact, prompted the authors *(118)* to investigate the antifeedant activity of its *trans*-fused analog (**18.27**). It was found that all *trans*-fused lactarane sesquiterpenes possessed decreased activity *(118)*. Isovellerol (**7.11**) a hydroxyaldehyde with a marasmane skeleton exhibited very high antifeedant activity *(80)*, and the suggestion was made that the high activity was caused by the fact that in solvents isovellerol can exist in three forms *(7)*, two cyclic hemiacetals, and the open structure (see Scheme 3). An isolactarane analog (**9.13**) (see Scheme 4) of isovellerol prepared by reduction of isolactarorufin (**9.4**) showed only small antifeedant activity which may due to the fact that it exists in a cyclic form stabilized by internal hydrogen bonding *(80)*.

References

1. Whittaker, R.H.: New Concepts of Kingdoms of Organisms. Science, **163**, 150 (1969).

2. Bon, M.: Clé Monographique du Genre *Lactarius* (Pers. ex Fr.) S.F. Gray. Documents Mycologiques, Tome X, Fascicule n. 40, Groupe de Mycologie Fondamentale et Appliquée, Lille, **10**, 1 (1980).

3. Gamba-Invernizzi, A., L. Garlaschelli, A. Rossi, G. Vidari, and P. Vita-Finzi: New Farnesane Sesquiterpenes from *Lactarius porninsis*. J. Nat. Prod., **56**, 1948 (1993).

4. Vokac, K., Z. Samek, V. Herout, and F. Sorm: On Terpenes. CCV. The Structure of Two Native Orange Substances from *Lactarius deliciosus* L. Coll. Czech, Chem. Commun., **35**, 745 (1970).

5. Bergendorff, O., and O. Sterner: The Sesquiterpenes of *Lactarius deliciosus* and *Lactarius deterrimus*. Phytochemistry, **27**, 97 (1988).

6. Harmon, A.D., K.H. Weisgraber, and U. Weiss: Preformed Azulene Pigments of *Lactarius indigo* (Schw.) Fries (Russulaceae, Basidiomycetes). Experientia, **36**, 54 (1980).

7. Sterner, O., R. Bergman, J. Kihlberg, and B. Wickberg: The Sesquiterpenes of *Lactarius vellereus* and their Role in a Proposed Chemical Defence System. J. Nat. Prod., **48**, 279 (1985).

8. GLUCHOFF-FIASSON, K., and R. KÜHNER: Le Principe Responsable du Bleuissement au Reactif Sulfovanillique des Cystides ou Lacticiferes de Divers *Homobasidiomycetes*: Interêt Taxonomique. C. R. Acad. Sci. Ser. III, **294**, 1067 (1982).
9. DE BERNARDI, M., L. GARLASCHELLI, L. TOMA, G. VIDARI, and P. VITA-FINZI: The Chemical Basis of Hot-tasting and Yellowing of the Mushrooms *Lactarius chrysorrheus* and *L. scrobiculatus*. Tetrahedron, **49**, 1489 (1993).
10. MAGNUSSON, G., S. THOREN, J. DAHMEN, and K. LEANDER: Fungal Extractives. VIII. Two Sesquiterpene Furans from *Lactarius*. Acta Chem. Scand., **B 28**, 841 (1974).
11. BATTAGLIA, R., M. DE BERNARDI, G. FRONZA, G. MELLERIO, G. VIDARI, and P. VITA-FINZI: Fungal Metabolites. VIII. Structures of New Sesquiterpenes from *Lactarius scrobiculatus*. J. Nat. Prod., **43**, 319 (1980).
12. STERNER, O., R. BERGMAN, J. KIHLBERG, J. OLUWADIYA, B. WICKBERG, G. VIDARI, M. DE BERNARDI, F. DE MARCHI, G. FRONZA, and P. VITA-FINZI: Basidiomycete Sesquiterpenes: The Silica Gel Induced Degradation of Velutinal Derivatives. J. Org. Chem., **50**, 950 (1985).
13. AYER, W.A., and L.M. BROWNE: Terpenoid Metabolites of Mushrooms and Related Basidiomycetes. Tetrahedron, **37**, 2199 (1981).
14. TURNER, W.: Fungal Metabolites. Academic Press, London (1971).
15. TURNER, W. and D. ALDRIDGE: Fungal Metabolites II. Academic Press, London (1983).
16. VIDARI, G., and P. VITA-FINZI: Sesquiterpenes and Other Secondary Metabolites of Genus *Lactarius* (Basidiomycetes): Chemistry and Biological Activity. Studies in Natural Products Chemistry, **17**, 152 (1995).
17. HANSSON, T., and O. STERNER: Studies of the Conversion of Sesquiterpenes in Injured Fruit Bodies of *Lactarius vellereus*. A Biomimetic Transformation of Stearoylvelutinal to Isovelleral. Tetrahedron Letters, **32**, 2541 (1991).
18. HANSSON, T., Z. PANG, and O. STERNER: The Conversion of [12-$^{2}H_{3}$]-Labelled Velutinal in Injured Fruit Bodies of *Lactarius vellereus*. Further Insight into the Biosynthesis of the Russulaceae Sesquiterpenes. Acta Chem. Scand., **47**, 403 (1993).
19. DANIEWSKI, W.M., P.A. GRIECO, J.C. HUFFMAN, A. RYMKIEWICZ, and A. WAWRZUŃ: 12-Hydroxycaryophyllene-4,5-oxide, a Sesquiterpene from *Lactarius camphoratus*. Phytochemistry, **20**, 2733 (1981).
20. DE BERNARDI, M., G. MELLERIO, G. VIDARI, P. VITA-FINZI, and G. FRONZA: Fungal Metabolites. Part 5. Uvidins, New Drimane Sesquiterpenes from *Lactarius uvidus* Fries. J. Chem. Soc. Perkin Trans. I, 221 (1980).
21. DE BERNARDI, M., G. MELLERIO, G. VIDARI, P. VITA-FINZI, and G. FRONZA: Fungal Metabolites. Part 15. Structure and Chemical Correlations of Uvidin C, D, and E, New Drimane Sesquiterpenes from *Lactarius uvidus* Fries. J. Chem. Soc. Perkin Trans. I, 2739 (1983).
22. GARLASCHELLI, L., G. MELLERIO, G. VIDARI, and P. VITA-FINZI: New Fatty Acid Esters of Drimane Sesquiterpenes from *Lactarius uvidus*. J. Nat. Prod., **57**, 905 (1994).
23. GARLASCHELLI, L., and G. VIDARI: Synthetic Studies on Biologically Active Natural Compounds. Part I: Stereospecific Transformation of Uvidin A into (−)-Cinnamodial. Tetrahedron, **45**, 7371 (1989).
24. GARLASCHELLI, L., P. DE TULLIO, and G. VIDARI: Synthetic Studies on Biologically Active Natural Compounds. Part III. Stereospecific Transformation of Uvidin A into (−)-Cinnamosmolide. Tetrahedron, **47**, 6769 (1991).
25. GILL, M., and W. STEGLICH: Pigments of Fungi (Macromycetes). In: Progress in the Chemistry of Organic Natural Products, Vol. 51 (W. HERZ, G.W. KIRBY, R.E. MOORE, W. STEGLICH, and CH. TAMM, eds.), p. 1. (Springer-Verlag, Wien-New York).

26. STERNER, O., O. BERGENDORFF, and F. BOCCHIO: The Isolation of a Guaiane Sesquiterpene from Fruit Bodies of *Lactarius sanguifluus*. Phytochemistry, **28**, 2501 (1989).
27. KOUL, S.K., S.C. TANEYA, S.P. IBRAHAM, K.L. DHAR, and C.K. ATAL: A C-formylated Azulene from *Lactarius deterrimus*. Phytochemistry, **24**, 181 (1985).
28. BERTELLI, D.J., and J.H. CRABTREE: Naturally Occurring Fulvene Hydrocarbons. Tetrahedron, **24**, 2079 (1968).
29. AYER, W.A., and L. TRIFONOV: Aromatic Compounds from Liquid Cultures of *Lactarius deliciosus*. J. Nat. Prod., **57**, 839 (1994).
30. SCHMITT, J.A.: Chemotaxonomic Morphological and Phytosociological Studies on Central European *Lactarius* Species Section Dapetes. Z. Pilzkunde, **39**, 219 (1973).
31. HEILBRONNER, E., and R.W. SCHMID: Zur Kenntnis der Sesquiterpene und Azulene. Azulenealdehyde und Azulenketone: Die Struktur des Lactaroviolins. Helv. Chim. Acta, **37**, 2018 (1954).
32. SORM, F., V. BENESOVA, J. KRUPICKA, V. SNEBERG, L. DOLEJS, V. HEROUT, and J. SICHER: On Terpenes. LXV. The Constitution of Lactaroviolin. Synthesis of 1-Ethyl-4-methyl-7-isopropylazulene and 4-Ethyl-1-methyl-7-isopropylazulene. Coll. Czech, Chem. Commun., **20**, 227 (1955).
33. BENESOVA, V., V. HEROUT, and F. SORM: Plant Substances. III. Substances from *Lactarius deliciosus* L. Coll. Czech, Chem. Commun., **19**, 1351 (1954).
34. DE ROSA, S., and S. DE STEFANO: Guaiane Sesquiterpenes from *Lactarius sanguifluus*. Phytochemistry, **26**, 2007 (1987).
35. VIDARI, G., L. GARLASCHELLI, A. ROSSI, and P. VITA-FINZI: New Protoilludane Sesquiterpenes from *Lactarius violascens*. Tetrahedron Letters, **39**, 1957 (1998).
36. FAVRE-BONVIN, J., K. GLUCHOFF-FIASSON, and J. BERNILLON: Structure du Stearyl-Velutinal, Sesquiterpenoide Naturel de *Lactarius velutinus* Bert. Tetrahedron Letters, **23**, 1907 (1982).
37. STERNER, O., R. BERGMAN, E. KESLER, L. NILSSON, J. OLUWADIYA, and B. WICKBERG: Velutinal Esters of *Lactarius vellereus* and *Lactarius necator*. The Preparation of Free Velutinal. Tetrahedron Letters, **24**, 1415 (1983).
38. STERNER, O.: The Russulaceae Sesquiterpenes. Ph.D. Dissertation. University of Lund, Sweden (1985).
39. STEGLICH, W., and O. STERNER: Isolierung von Sesquiterpenoiden aus der Becher-noralle. *Artomyces pyxidatus* (Clavicoronaceae). Z. Mykol, **54**, 175 (1988).
40. CAMAZINE, S., and A.T.J. LUPO: Labile Toxic Compounds of the Lactarii: the Role of the Laticiferous Hyphae as a Storage Depot for Precursors of Pungent Dialdehydes. Mycologia, **76**, 355 (1984).
41. CAMAZINE, S., J.F. RESCH, T. EISNER, and J. MEINWALD: Mushroom Chemical Defense: Pungent Sesquiterpenoid Dialdehyde Antifeedant to Opossum. J. Chem. Ecol., **23**, 1439 (1983).
42. HANSSON, T., O. STERNER, and A. STRID: Chemotaxonomic Evidence for a Division of *Lactarius vellereus* and *Lactarius bertillonii* as Different Species. Phytochemistry, **39**, 363 (1995).
43. DANIEWSKI, W.M., M. GUMULKA, K. PTASZYŃSKA, P. SKIBICKI, J. KRAJEWSKI, and P. GLUZIŃSKI: Marasmane Lactones from *Lactarius vellereus*. Phytochemistry, **31**, 913 (1992).
44. DANIEWSKI, W.M., M. KOCÓR, T. JANUSZEWSKI, and A. RYMKIEWICZ: Constituents of Higher Fungi. Part XI. New Monohydroxylactone from *Lactarius necator*. Sesquiterpene Monohydroxylactone Contents as Characteristic Chemotaxonomic Features of Various *Lactarius* Species. Polish J. Chem., **55**, 807 (1981).

45. TALVITIE, A., K.-G. WIDEN, and E.-L. SEPPA: ^{1}H NMR Spectroscopic Study of Blennin A and 15-Hydroxyblennin A, Two Sesquiterpene Lactones from *Lactarius torminosus* (*Russulaceae*) Mushrooms. Finn. Chem. Lett., 62 (1980).
46. PYYSALO, H.: Identification of Volatile Compounds in Seven Edible Fresh Mushrooms. Acta Chem. Scand., B, **30**, 235 (1976).
47. STERNER, O.: The Co-formation of Sesquiterpene Aldehydes and Lactones in Injured Fruit Bodies of *Lactarius necator* and *L. circellatus*. The Isolation of *epi*-Piperalol. Acta Chem. Scand., **43**, 694 (1989).
48. DANIEWSKI, W.M., M. GUMULKA, K. PTASZYŃSKA, G. VIDARI, L. GARLASCHELLI, G. FRONZA, and M. BUDESINSKY: 7-*Epi*-pipertriol, a Lactarane Sesquiterpene from *Lactarius necator*. Phytochemistry, **27**, 3314 (1988).
49. PANG, Z., F. BOCCHIO, and O. STERNER: The Isolation of New Sesquiterpene Aldehydes from Injured Fruit Bodies of *Lactarius scrobiculatus*. Tetrahedron Letters, **33**, 6863 (1992).
50. DE BERNARDI, M., G. FRONZA, G. VIDARI, and P. VITA-FINZI: Fungal Metabolites II: New Sesquiterpenes from *Lactarius scrobiculatus* Scop. (*Russulaceae*). Chim. e Ind., **58**, 177 (1976).
51. CRADWICK, P.D., and G.A. SIM: Crystallographic Determinations of Partial Stereochemistries of the Sesquiterpenoids Illudol and Marasmic Acid. J. Chem. Soc., Chem. Commun., 431 (1971).
52. COMER, F.W., F. MCCAPRA, I.H. QURESHI, and A.I. SCOTT: The Structure and Chemistry of Hirsutic Acid. Tetrahedron, **23**, 4761 (1967).
53. LIST, P.H., and H. HACKENBERG: Velleral und iso-Velleral. Scharf Schmeckende Stoffe aus *Lactarius vellereus* Fries. Arch. Pharmaz., **302**, 125 (1969).
54. MAGNUSSON, G., S. THOREN, and B. WICKBERG: Fungal Extractives I. Structure of a Sesquiterpene Dialdehyde from *Lactarius* by Computer Simulation of the NMR Spectrum. Tetrahedron Letters, 1105 (1972).
55. MAGNUSSON, G., S. THOREN, and T. DRAKENBERG: Fungal Extractives-IV. Structure of a Novel Sesquiterpene Dialdehyde from *Lactarius* by Spectroscopic Methods. Tetrahedron, **29**, 1621 (1973).
56. DE BERNARDI, M., G. FRONZA, G. MELLERIO, V. VALLA, G. VIDARI, and P. VITA-FINZI: Fungal Metabolites. XVII. Sesquiterpenes from *Lactarius pallidus* Persoon. Gazz. Chim. Ital., **114**, 163 (1984).
57. BERGMAN, R., T. HANSSON, O. STERNER, and B. WICKBERG: A Total Synthesis of (+)-Isovelleral. The Absolute Configuration of the Russulaceae Sesquiterpenes. J. Chem. Soc., Chem. Commun., 865 (1990).
58. DANIEWSKI, W.M., W. KROSZCZYŃSKI, P. SKIBICKI, M. DE BERNARDI, G. FRONZA, G. VIDARI, and P. VITA-FINZI: Normarasmane Sesquiterpenes from *Lactarius vellereus*. Phytochemistry, **27**, 187 (1988).
59. HANSSON, T., O. STERNER, B. WICKBERG, and R. BERGMAN: The Thermal Isomerization of the Sesquiterpene Isovelleral and Merulidial. A Reversible Ring Opening of the *cis*-Methylcyclopropanecarbaldehyde Group via an Intramolecular Ene Reaction. J. Org. Chem., **57**, 3822 (1992).
60. STERNER, O., R. BERGMAN, C. FRANZEN, and B. WICKBERG: New Sesquiterpenes in a Proposed Russulaceae Chemical Defense System. Tetrahedron Letters, **26**, 3163 (1985).
61. STERNER, O., R.E. CARTER, and L.M. NILSSON: Structure-Activity Relationship for Unsaturated Dialdehydes 1. The Mutagenic Activity of 18 Compounds in the Salmonella / Microsome Assay. Mutation Res., **188**, 169 (1987).

62. Jonassohn, M., H. Anke, P. Morales, and O. Sterner: Structure — Activity Relationship for Unsaturated Dialdehydes. The Generation of Bioactive Products by Autoxidation of Isovelleral and Merulidial. Acta Chem. Scand., **49**, 530 (1995).
63. Daniewski, W.M., M. Gumułka, K. Ptaszyńska, P. Skibicki, G. Fronza, and G. Vidari: Constituents of Higher Fungi. XX. New Sesquiterpenoid Triol of Marasmane Skeleton from *Lactarius vellereus*. Bull. Acad. Polon. Sci., Ser. Sci. Chim., **37**, 283 (1989).
64. Daniewski, W.M., M. Gumułka, P. Skibicki, U. Jacobsson, and T. Norin: Constituents of Higher Fungi. Part XIX. New Sesquiterpenoid Lactone of Marasmane Skeleton from *Lactarius vellereus*. Bull. Acad. Polon. Sci., Ser. Sci. Chim., **35**, 251 (1987).
65. Nozoe, S., H. Matsumoto, and S. Urano: The Structure of New Sesquiterpenes from Basidiomycetes. Tetrahedron Letters, 3125 (1971).
66. Daniewski, W.M., M. Gumułka, D. Truszewska, U. Jacobsson, and T. Norin: Monohydroxylactones of *Lactarius vellereus*. Phytochemistry, **41**, 1093 (1996).
67. Daniewski, W.M., M. Gumułka, P. Skibicki, W. Anczewski, U. Jacobsson, and T. Norin: New Constituents of *Lactarius vellereus*. Natural Product Letters, **5**, 123 (1994).
68. De Bernardi, M., G. Vidari, P. Vita-Finzi, and K. Gluchoff-Fiasson: Biogenesis-Like Conversion of Marasmane to Lactarane and Seco-Lactarane Skeleton. Tetrahedron Letters, **23**, 4623 (1982).
69. De Bernardi, M., G. Fronza, A.P. Gatti, G. Vidari, and P. Vita-Finzi: New Marasmane Sesquiterpenes from *Lactarius rubrocinctus* (Basidiomycetes). 15th IUPAC International Symposium on the Chemistry of Natural Products, PA 75 (1986).
70. Abreu, P., M. Araujo, T. Fonseca, and S.P. Santos: Chemical Composition of *Lactarius controversus*, Basidiomycetes Fungus. Pharm. Pharmacol. Lett., **7**, 138 (1997).
71. Nozoe, S., H. Kobayashi, S. Urano and J. Furukawa: Isolation of Δ^6-Protoilludene and the Related Alcohols. Tetrahedron Letters, 1381 (1977).
72. Sterner, O., R. Bergman, and B. Wickberg: The Synthetic Preparation of the Fungal Sesquiterpene Velutinal from Related Compounds. Finn. Chem. Lett., 116 (1985).
73. Gluziński, P., W.M. Daniewski, M. Gumułka, and D. Przesmycka: Crystal and Molecular Structure of 10-α-(N-trichloroacetyl-carbamate)-13-hydroxy-marasm-7-en-5-oic Acid γ-Lactone. Polish J. Chem., **70**, 458 (1996).
74. Favre-Bonvin, J., and K. Gluchoff-Fiasson: Structures of Two Glutinopallal Esters, New Natural Sesquiterpenoids from *Lactarius glutinopallens*. Phytochemistry, **27**, 286 (1988).
75. Daniewski, W.M., and M. Kocór: Isolation and Structure of Some New Sesquiterpenes from *Lactarius rufus*. Bull. Acad. Polon. Sci., Ser. Sci. Chim., **18**, 585 (1970).
76. Daniewski, W.M., M. Kocór, and S. Thoren: Constituents of Higher Fungi. Part VIII. Isolactarorufin, a Novel Tetracyclic Sesquiterpene Lactone from *Lactarius rufus*. Heterocycles, **5**, 77 (1976).
77. Konitz, A., M. Bogucka-Ledóchowska, Z. Dauter, A. Hempel, and E. Borowski: The Structure of Isolactarorufin. Tetrahedron Letters, 3401 (1977).
78. Daniewski, W.M., M. Kocór, and S. Thoren: Constituents of Higher Fungi. Part X. Isolactarorufin, a Novel Tetracyclic Sesquiterpene Lactone from *Lactarius rufus*. Polish J. Chem., **52**, 561 (1978).

79. DANIEWSKI, W.M., W. KROSZCZYŃSKI, and A. WAWRZUŃ: Constituents of Higher Fungi. Part XVIII. Dihydroxylactone Contents of Various *Lactarius* Species as Characteristic Chemotaxonomic Features. Polish J. Chem., **61**, 123 (1987).
80. DANIEWSKI, W.M., M. GUMUŁKA, E. BŁOSZYK, U. JACOBSSON, and T. NORIN: Isovellerol and New Isolactarane Sesquiterpenes, Their Structure and Antifeedant Activity. Polish J. Chem., **71**, 1254 (1997).
81. FROBORG, J., and G. MAGNUSSON: Fungal Extractives. 12. Construction of the Vellerane Skeleton with Total Syntheses of Racemic Velleral, Vellerolactone, and Pyrovellerolactone. Revised Structures. J. Amer. Chem. Soc., **100**, 6728 (1978).
82. MAGNUSSON, G., and S. THOREN: Fungal Extractives V. The Stereostructure of two Sesquiterpene Lactones from *Lactarius*. Acta Chem. Scand., **27**, 2396 (1973).
83. DANIEWSKI, W.M., and M. KOCÓR: Constituents of Higher Fungi. II. Structure of Lactarorufin A. Bull. Acad. Polon. Sci., Ser. Sci. Chim., **19**, 553 (1971).
84. DE BERNARDI, M., G. FRONZA, G. MELLERIO, G. VIDARI, and P. VITA-FINZI: New Sesquiterpene Hydroxylactones from *Lactarius* Species. Phytochemistry, **18**, 293 (1979).
85. DANIEWSKI, W.M., M. KOCÓR, and J. KRÓL: Constituents of Higher Fungi. Part VII. Lactarorufin N and Revised Structures of Lactarorufins. Rocz. Chem., **50**, 2095 (1976).
86. DANIEWSKI, W.M., A. WAWRZUŃ, M. DE BERNARDI, G. VIDARI, P. VITA-FINZI, G. FRONZA, and G. GATTI: Structural Studies on *Lactarius* Sesquiterpenes: Structure Elucidation of Lactarorufins D and E and Conformational Analysis of Lactaran-5-olides. Tetrahedron, **40**, 2757 (1984).
87. DANIEWSKI, W.M., M. GUMUŁKA, K. PTASZYŃSKA, P. SKIBICKI, U. JACOBSSON, and T. NORIN: Synthesis of 5-Hydroxy-lactar-6-en-13-oic-acid γ-Lactones, the Sesquiterpenoid Derivatives of *Lactarius* Origin. Polish J. Chem., **66**, 791 (1992).
88. ZHANG, J., and X.Z. FENG: Sesquiterpene Hydroxylactones from *Lactarius subvellereus*. Phytochemistry, **46**, 157 (1997).
89. DANIEWSKI, W.M., A. EJCHART, J. JURCZAK, L. KOZERSKI, and J.S. PYREK: Constituents of Higher Fungi. Part III. Confirmation and Stereochemistry of Structure of Lactarorufin A by $Eu(dpm)_3$ Shifted NMR Spectra. Bull. Acad. Polon. Sci., Ser. Sci. Chim., **20**, 131 (1972).
90. GARLASCHELLI, L., L. TOMA, G. VIDARI, and D. COLOMBO: Conformational Studies and Stereochemical Assignments of the Lactarane Sesquiterpenes Furoscrobiculin D and Blennin D. Tetrahedron, **50**, 1211 (1994).
91. DANIEWSKI, W.M., M. GUMUŁKA, K. PTASZYŃSKA, B. KAMIEŃSKI, P. SKIBICKI, U. JACOBSSON, and T. NORIN: Constituents of Higher Fungi. XXI. Transformation of Lactarorufin A into Lactarorufin E and the Synthesis of 3-*Epi*-lactarorufin D. Bull. Acad. Polon. Sci., Ser. Sci. Chim., **37**, 289 (1989).
92. BOSETTI, A., G. FRONZA, G. VIDARI, and P. VITA-FINZI: Norlactarane and Lactarane Sesquiterpenes from *Lactarius scrobiculatus*. Phytochemistry, **28**, 1427 (1989).
93. DANIEWSKI, W.M., M. GUMUŁKA, P. SKIBICKI, J. KRAJEWSKI, and P. GLUZIŃSKI: 2(3)-8(9)-Bisanhydrolactarorufin A and a Highly Oxygenated Furanol from *Lactarius vellereus*. Phytochemistry, **30**, 1326 (1991).
94. DANIEWSKI, W.M., M. GUMUŁKA, K. PTASZYŃSKA, P. SKIBICKI, U. JACOBSSON, and T. NORIN: 3-Deoxy-3-*Epi*-lactaroscrobiculide B, a Sesquiterpene from *Lactarius vellereus*. Phytochemistry, **31**, 3933 (1992).
95. DANIEWSKI, W.M., M. KOCÓR, and B. ŻÓŁTOWSKA: Constituents of Higher Fungi. Part V. Structure of Lactarorufin B. Bull. Acad. Polon. Sci., Ser. Sci. Chim., **21**, 785 (1973).

96. MAGNUSSON, G., and S. THOREN: Fungal Extractives III. Two Sesquiterpene Lactones from *Lactarius*. Acta Chem. Scand., **27**, 1573 (1973).

97. DANIEWSKI, W.M., M. GUMUŁKA, E. PANKOWSKA, K. PTASZYŃSKA, E. BŁOSZYK, U. JACOBSSON, and T. NORIN: 3,8-Ethers of Lactarane Sesquiterpenes. Phytochemistry, **32**, 1499 (1993).

98. DE BERNARDI, M., G. FRONZA, G. VIDARI, and P. VITA-FINZI: Fungal Metabolites XX: Chemical Correlation of Lactarane and Secolactarane Sesquiterpenes. Absolute Configuration of Furosardonin A, Lactaral and Blennin C. Tetrahedron, **42**, 4277 (1986).

99. DANIEWSKI, W.M., M. KOCÓR, and J. KRÓL: Constituents of Higher Fungi. VI. Two New Sesquiterpenoic Lactones from *Lactarius necator*. Bull. Acad. Polon. Sci., Ser. Sci. Chim., **23**, 637 (1975).

100. DANIEWSKI, W.M., W. KROSZCZYŃSKI, A. WAWRZUŃ, and A. RYMKIEWICZ: Constituents of Higher Fungi. Part XVI. Identification of *Lactarius* Species by HPLC Using Sesquiterpene Monohydroxylactone Contents as Characteristic Chemotaxonomic Features. J. Liquid Chrom., **7**, 2915 (1984).

101. SEPPA, E.-L., and K.-G. WIDEN: Sesquiterpenoids of *Lactarius torminosus* and *Lactarius trivialis* and Correlation of the Sesquiterpenoid Composition of *Lactarius* Species with the Generic Subdivision. Ann.Bot.Fennici, **17**, 56 (1980).

102. PYYSALO, H., E.-L. SEPPA, and K.-G. WIDEN: Application of Gas Chromatography to the Analysis of Sesquiterpene Lactones from *Lactarius* (*Russulaceae*) Mushrooms. J. Chrom., **190**, 466 (1980).

103. VIDARI, G., M. DE BERNARDI, P. VITA-FINZI, and G. FRONZA: Sesquiterpenes from *Lactarius blennius*. Phytochemistry, **15**, 1953 (1976).

104. DANIEWSKI, W.M., and J. KRÓL: Constituents of Higher Fungi. Part XII. Rearrangement of Lactone Ring of 3-Deoxy-6β,7β-dihydro-8-epilactarorufin A. Corrected Structures of Lactarorufin N and 3-Deoxylactarorufin A. Polish J. Chem., **55**, 1247 (1981).

105. DE BERNARDI, M., G. FRONZA, G. VIDARI, and P. VITA-FINZI: Stereochemistry of Blennin A and Blennin D from *Lactarius blennius*. Phytochemistry, **19**, 99 (1980).

106. DANIEWSKI, W.M., M. KOCÓR, and J. KRÓL: Constituents of Higher Fungi. Part IX. Structure of 3-Deoxylactarorufin A. Rocz. Chem., **51**, 1395 (1977).

107. BOGUCKA-LEDÓCHOWSKA, M., A. HEMPEL, Z. DAUTER, A. KONITZ, and E. BOROWSKI: The Structure of Lactarorufin B -3,8-ether 14-*p*-bromobenzoate. Tetrahedron Letters, 2267 (1976).

108. BARANOWSKA, E., and W.M. DANIEWSKI: Constituents of Higher Fungi. IV. Mass Spectroscopic Investigations of Lactarorufin A and Its Derivatives. Bull. Acad. Polon. Sci., Ser. Sci. Chim., **20**, 313 (1972).

109. DANIEWSKI, W.M., M. GUMUŁKA, K. PTASZYŃSKA, J. SITKOWSKI, P. SKIBICKI, U. JACOBSSON, and T. NORIN: Constituents of Higher Fungi. XXIII. New Lactarolide from *Lactarius mitissimus*. Bull. Acad. Polon. Sci., Ser. Sci. Chim., **39**, 251 (1991).

110. DANIEWSKI, W.M., M. GUMUŁKA, and P. SKIBICKI: Furantriol, a Lactarane Sesquiterpene from *Lactarius mitissimus*. Phytochemistry, **29**, 527 (1990).

111. DANIEWSKI, W.M., P. GLUZIŃSKI, M. GUMUŁKA, J.W. KRAJEWSKI, and K. PTASZYŃSKA: Synthesis of 9-*epi*-(*trans*-fused) Lactarane Sesquiterpenes. Polish J. Chem., **68**, 287 (1994).

112. ANDINA, D., M. DE BERNARDI, A. DEL VECCHIO, G. FRONZA, G. MELLERIO, G. VIDARI, and P. VITA-FINZI: Sesquiterpenes from *Russula sardonia*. Phytochemistry, **19**, 93 (1980).

113. WIDEN, K.-G., and E.-L. SEPPA: 15-Hydroxyblennin A, a New Lactarane-Type Sesquiterpene Lactone Isolated from *Lactarius torminosus*. Phytochemistry, **18**, 1226 (1979).

114. DANIEWSKI, W.M., W. KROSZCZYŃSKI, and J. KRÓL: Constituents of Higher Fungi. Part XIV. Regioselectivity of Furan Oxidation Induced by Neighboring Hydroxyl Groups. Transformation of Furandiol into Lactarorufin A. Polish J. Chem., **57**, 483 (1983).

115. DANIEWSKI, W.M., M. GUMUŁKA, K. PTASZYŃSKA, P. SKIBICKI, E. BŁOSZYK, B. DROŻDŻ, S. STROMBERG, T. NORIN, and M. HOLUB: Antifeedant Activity of Some Sesquiterpenoids of the Genus *Lactarius*. Eur. J. Entomol., **90**, 65 (1993).

116. DANIEWSKI, W.M., M. GUMUŁKA, P. GLUZIŃSKI, J. KRAJEWSKI, E. PANKOWSKA, K. PTASZYŃSKA, J. SITKOWSKI, and E. BŁOSZYK: 3-Ethoxy Lactarane Sesquiterpenes of *Lactarius* Origin Antifeedant Activity. Polish J. Chem., **66**, 1249 (1992).

117. ZHANG, J., and X.Z. FENG: Subvellerolactone C, a New Lactarane Sesquiterpene from *Lactarius subvellereus*. Chin. Chem. Lett., **7**, 1097 (1996).

118. DANIEWSKI, W.M., M. GUMUŁKA, W. ANCZEWSKI, E. BŁOSZYK, B. DROŻDŻ, U. JACOBSSON, and T. NORIN: 3-Ethoxy-9-*epi*-(*trans*-fused) Lactarane Sesquiterpenes of *Lactarius* Origin. Antifeedant Activity. Polish J. Chem., **69**, 1687 (1995).

119. VIDARI, G., L. GARLASCHELLI, M. DE BERNARDI, G. FRONZA, and P. VITA-FINZI: The Structure of a New Epoxylactone from *Lactarius scrobiculatus* Scop. (*Russulaceae*) by Spectroscopic Methods. Tetrahedron Letters, 1773 (1975).

120. STERNER, O., O. WIK, and R.E. CARTER: The Structure of a Novel Fungal Sesquiterpene, Elucidated by Spectral Methods. Acta Chem. Scand., **B 42**, 43 (1988).

121. DANIEWSKI, W.M., P. GLUZIŃSKI, J.W. KRAJEWSKI, and P. SKIBICKI: X-ray Structural Investigation of 5,13-Epoxy-3β-hydroxy-lactara-2(9),5,7(13)-trien-4,8-dion. J. Cryst. Spectr. Research, **21**, 407 (1991).

122. FROBORG, J., and G. MAGNUSSON: Fungal Extractives-XI. On the Biogenetic Interrelationship between the Marasmane- and Vellerane Sesquiterpene Skeletons. Tetrahedron, **34**, 2027 (1978).

123. GARNIER, J., J. MAHUTEAU, and M. PLAT: Constituents of *Lactarius hepaticus*. Plant. Med. Phytother., **24**, 87 (1990).

124. KIHLBERG, J., R. BERGMAN, L. NILSSON, O. STERNER, and B. WICKBERG: The Structure of a Novel Sesquiterpene Furan Alcohol with a Lactarane Skeleton. Tetrahedron Letters, **24**, 4631 (1983).

125. KRAJEWSKI, J.W., P. GLUZIŃSKI, W.M. DANIEWSKI, M. GUMUŁKA, K. PTASZYŃSKA, A. KEMME, and A. MISHNEV: Crystal and Molecular Structure of 8-*Epi*-9-*epi*-furandiol, an Unusual *Trans*-fused Lactarane Sesquiterpene. Polish J. Chem., **68**, 515 (1994).

126. GLUZIŃSKI, P., W.M. DANIEWSKI, and M. GUMUŁKA: Crystal and Molecular Structure of 3-O-Ethyl-8-*epi*-9-*epi*-furandiol, a *Trans*-fused Lactarane Sesquiterpene. Polish J. Chem., **70**, 60 (1996).

127. MAGNUSSON, G., and S. THOREN: Fungal Extractives-VI. Structure of Lactaral, a New Sesquiterpene Furan-3-aldehyde from *Lactarius*, Spectroscopic Methods. Tetrahedron, **30**, 1431 (1974).

128. FROBORG, J., G. MAGNUSSON, and S. THOREN: Fungal Extractives. VII. A Formal Synthesis of (±)-Lactaral. Acta Chem. Scand., **B 28**, 265 (1974).

129. SUORTTI, T., A. WRIGHT, and A. KOSKINEN: Necatorin, a Highly Mutagenic Compound from *Lactarius necator*. Phytochemistry, **22**, 2873 (1983).

130. HILGER, C.S., B. FUGMANN, and W. STEGLICH: Synthesis of Necatorone. Tetrahedron Letters, **26**, 5975 (1985).

131. FUGMANN, B., B. STEFFAN, and W. STEGLICH: Necatorone, an Alkaloidal Pigment from the Gilled Toadstool *Lactarius necator* (Agaricales). Tetrahedron Letters, **25**, 3575 (1984).

132. KLAMANN, J.-D., B. FUGMANN, and W. STEGLICH: Alkaloidal Pigments from *Lactarius necator* and *L. atroviridis*. Phytochemistry, **28**, 3519 (1989).

133. TAKAHASHI, A., G. KUSANO, T. OHTA, and S. NOZOE: The Constituents of *Lactarius flavidulus* Imai. Chem. Pharm. Bull., **36**, 2366 (1988).

134. FUJIMOTO, H., Y. NAKAYAMA, and M. YAMAZAKI: Identification of Immunosuppressive Components of a Mushroom, *Lactarius flavidulus*. Chem. Pharm. Bull., **41**, 654 (1993).

135. TAKAHASHI, A., G. KUSANO, T. OHTA, and S. NOZOE: Revised Structures of Flavidulols. Constituents of *Lactarius flavidulus* Imai and the Structure of Flavidulol D. Chem. Pharm. Bull., **41**, 2032 (1993).

136. GOMEZ, F., L. QUIJANO, J.S. CALDERON, and T. RIOS: Terpenoids Isolated from *Wigandia kunthii*. Phytochemistry, **19**, 2202 (1980).

137. VIDARI, G., P. VITA-FINZI, A.M. ZANOCCHI, and G. PEDRALI NOY: A Bioactive Tetraprenylphenol from *Lactarius lignyotus*. J. Nat. Prod., **58**, 893 (1995).

138. CIMINO, G., S. DE STEFANO, and L. MINALE: Prenylated Quinones in Marine Sponges: *Ircinia sp.* Experientia, **28**, 1401 (1972).

139. REYNOLDS, G.W.: Prenylated Hydroquinones: Contact Allergenes from Trichomes of *Phacelia minor* and *P. parryi*. Phytochemistry, **20**, 1365 (1981).

140. DE BERNARDI, M., G. VIDARI, P. VITA-FINZI, and G. FRONZA: The Chemistry of *Lactarius fuliginosus* and *Lactarius picinus*. Tetrahedron, **48**, 7331 (1992).

141. CONCA, E., M. DE BERNARDI, G. FRONZA, M.A. GIROMETTA, G. MELLERIO, G. VIDARI, and P. VITA-FINZI: Fungal Metabolites 10. New Chromenes from *Lactarius fuliginosus* Fries and *Lactarius picinus* Fries. Tetrahedron Letters, **22**, 4327 (1981).

142. ALLIEVI, C., M. DE BERNARDI, F. DEMARCHI, and G. MELLERIO: Chromatographic Analysis of 2,2-Dimethylchromene Derivatives. J. Chrom., **261**, 311 (1983).

143. ZHANG, J., and X.Z. FENG: Lactariolide, a New 14-Membered Ring Compound from *Lactarius subvellereus*. Chin. Chem. Lett., **8**, 135 (1997).

144. KOBATA, K., T. WADA, Y. HAYASHI, and H. SHIBATA: Studies on Chemical Components of Mushrooms. Part III. Volemolide, a Novel Norsterol from the Fungus *Lactarius volemus*. Biosci. Biotechnol. Biochem., **58**, 1542 (1994).

145. THOMPSON, S.K., and C.H. HEATHCOCK: Total Synthesis of Some Marasmane and Lactarane Sesquiterpenes. J. Org. Chem., **57**, 5979 (1992).

146. HANSSON, T., R. BERGMAN, O. STERNER, and B. WICKBERG: The Mechanism of the Thermal Rearrangement of the Marasmane Sesquiterpene (+) Isovelleral. Cyclopropane Ring Closure *via* an Intramolecular Ene Reaction. J. Chem. Soc., Chem. Commun., 1260 (1990).

147. FERLAND, J.M., Y. LEFEBVRE, R. DEGHENHI, and K. WIESNER: Synthetic New Cardenolides. Tetrahedron Letters, 3617 (1966).

148. ZIEGLER, F.E., and B.B. JAYNES: Rearrangement of a Hindered Allylic Alcohol During Vanadium-Catalyzed Epoxidation. A Short Synthesis of Uvidin C. Tetrahedron Letters, **26**, 5875 (1985).

149. LOPEZ, J., J. SIERRA, and M. CORTES: Synthesis of Natural Uvidin C. Chem. Lett., 2073 (1986).

150. THOMPSON, S.K., and C.H. HEATHCOCK: Total Synthesis of (±)-Isovelleral, a Mutagenic Sesquiterpene Dialdehyde from *Lactarius vellereus*. J. Org. Chem., **55**, 3004 (1990).

151. FEX, T., J. FROBORG, G. MAGNUSSON, and S. THOREN: Fungal Extractives. 10. An Alternative Synthesis of the Velleral Skeleton. J. Org. Chem., **41**, 3518 (1976).
152. TOCHTERMANN, W., S. BRUHN, M. MEINTS, and C. WOLFF: Synthese Funktionalisierte Hydroazulene, ein Neuer Zugang zum Lactaran-Gerüst. Tetrahedron, **50**, 9657 (1994).
153. TOCHTERMANN, W., S. BRUHN, M. MEINTS, C. WOLFF, E.-M. PETERS, K. PETERS, and H.G. VON SCHNERING: Synthese Stereoisomerer 4,7-Epoxy-Hydroazulen-5,6-Dicarbonsauredimethylester mit Lactaran-Gerüst. Tetrahedron, **51**, 1623 (1995).
154. WOCKENFUS, B., C. WOLFF, and W. TOCHTERMANN: Synthesis of 2(3)-8(9)-Bisanhydrolactarorufin A. Tetrahedron, **53**, 13703 (1997).
155. PRICE, M.E., and N.E. SCHORE: Total Synthesis of Furanether B. J. Org. Chem., **54**, 5662 (1989).
156. PRICE, M.E., and N.E. SCHORE: Efficient Synthetic Entry to Oxygen-Bridged Lactaranes Using Organometallic Methodology: A Short Synthesis of Furanether B. Tetrahedron Letters, **30**, 5865 (1989).
157. PRICE, M.E. and N.E. SCHORE: Improved Synthesis of 3,4-Disubstituted Furans: Use of Phase-Transfer Conditions. J. Org. Chem., **54**, 2777 (1989).
158. MOLANDER, G.A., and J.S. CAREY: Total Synthesis of Furanether B. An Application of a [3+4] Annulation Strategy. J. Org. Chem., **60**, 4845 (1995).
159. BELL, R.P.L., A. SOBOLEV, J.B.P.A. WIJNBERG, and A. DE GROOT: Base-Induced Rearrangement of Perhydronaphthalene-1,4-diol Monosulphonate Esters to 11-Oxatricyclo [5.3.1.$0^{2,6}$]undecanes. Total Synthesis of Furanether B. J. Org. Chem., **63**, 122 (1998).
160. OGINO, T., C. KURIHARA, Y. BABA, and K. KANEMATSU: Total Synthesis of Furoscrobiculin B. J. Chem. Soc., Chem. Commun., 1979 (1994).
161. SEKI, M., T. SAKAMOTO, H. SUEMUNE, and K. KANEMATSU: Total Synthesis of (±)-Furoscrobiculin B. J. Chem. Soc. Perkin Trans. I, 1707 (1997).
162. TANIS, S.P., and D.B. HEAD: Furans in Synthesis. The Preparation of (±)-Lactaral. Tetrahedron Letters, **23**, 5509 (1982).
163. NAWROT, J., E. BŁOSZYK, J. HARMATHA, L. NOVOTNY, and B. DROŻDŻ: Action of Antifeedants of Plant Origin on Beetles Infesting Stored Products. Acta Entomol. Bohemoslov., **83**, 327 (1986).
164. NAWROT, J., E. BŁOSZYK, H. GRABARCZYK, B. DROŻDŻ, W.M. DANIEWSKI, and M. HOLUB: Further Evaluation of Feeding Deterrence of Sesquiterpene Lactones to Storage Pests. Prace Naukowe. Pr. Nauk. IOR, **25**, 91 (1983).
165. KNUUTINEN, J., and A. WRIGHT: The Mutagenicity of *Lactarius* Mushrooms. Mutation Res., **103**, 115 (1982).
166. STERNER, O., R. BERGMAN, E. KESLER, G. MAGNUSSON, L. NILSSON, B. WICKBERG, and E. ZIMERSON: Mutagens in Larger Fungi, I. Forty-eight Species Screened for Mutagenic Activity in the Salmonella Microsome Assay. Mutation Res., **101**, 269 (1982).
167. ANKE, H., O. BERGENDORFF, and O. STERNER: Assays of the Biological Activities of Guaiane Sesquiterpenoids Isolated from Fruit Bodies of Edible *Lactarius* Species. Food Chem. Toxicol., **27**, 393 (1989).
168. ANKE, H., and O. STERNER: Comparison of the Antimicrobial and Cytotoxic Activities of Twenty Unsaturated Sesquiterpene Dialdehydes from Plants and Mushrooms. Planta Medica, **57**, 344 (1991).
169. GIANETTI, B., B. STEFFAN, and W. STEGLICH: Antibiotics from Basidiomycetes. Part 23. Merulidial, an Isolactarane Derivative from *Merulius tremellosus*. Tetrahedron, **42**, 3579 (1986).

170. DANIEWSKI, W.M., M. GUMUŁKA, D. PRZESMYCKA, K. PTASZYŃSKA, E.,. BŁOSZYK, and B. DROŻDŻ: Sesquiterpenes of *Lactarius* Origin, Antifeedant Structure-Activity Relationship. Phytochemistry, **38**, 1161 (1995).

171. FROBORG, J., G. MAGNUSSON, and S. THOREN: Fungal Extractives. IX. Synthesis of the Velleral Skeleton and a Total Synthesis of Pyrovellerolactone. J. Org. Chem., **40**, 1595 (1975).

172. GAWROŃSKI, J.K., A. OEVEREN, H. DEEN, C.W. LEUNG, and B.L. FERINGA: Simple Circular Dichroic Method for the Determination of Absolute Configuration of 5-Substituted 2(5*H*)-Furanones. J. Org. Chem., **61**, 1513 (1996).

(Received December 8, 1998)

Author Index

Page numbers printed in *italics* refer to References

Subject Index

Springer-Verlag and the Environment

FSC www.fsc.org
MIX
Papier aus verantwortungsvollen Quellen
Paper from responsible sources
FSC® C105338

If you have any concerns about our products,
you can contact us on
ProductSafety@springernature.com

In case Publisher is established outside the EU,
the EU authorized representative is:
Springer Nature Customer Service Center GmbH
Europaplatz 3, 69115 Heidelberg, Germany

Printed by Libri Plureos GmbH
in Hamburg, Germany